METAMATHEMATICS, MACHINES, AND GÖDEL'S PROOF

Cambridge Tracts in Theoretical Computer Science

Titles in the series

METAMATHEMATICS, MACHINES, AND GÖDEL'S PROOF

N. SHANKAR

SRI International

CAMBRIDGE
UNIVERSITY PRESS

PUBLISHED BY THE PRESS SYNDICATE OF THE UNIVERSITY OF CAMBRIDGE
The Pitt Building, Trumpington Street, Cambridge CB2 1RP, United Kingdom

CAMBRIDGE UNIVERSITY PRESS
The Edinburgh Building, Cambridge CB2 2RU, United Kingdom
40 West 20th Street, New York, NY 10011–4211, USA
10 Stamford Road, Oakleigh, Melbourne 3166, Australia

First published 1994
First paperback edition 1997

A catalogue record for this book is available from the British Library

Library of Congress Cataloguing in Publication data available

ISBN 0 521 58533 3 paperback

Transferred to digital printing 2002

To My Parents

Contents

List of Figures

Preface

A: It's true!
B: It's not!
A: Yes, it is!
B: It couldn't be!
A: It is true!
B: Prove it!
A: Oh, it can't be proved, but nevertheless it's true.
B: Now just a minute: How can you say it's true if it can't be proved?
A: Oh, there are certain things that are true even though they can't be proved.
B: That's not true!
A: Yes, it is; Gödel proved that there are certain things that are true but that cannot be proved.
B: That's not true!
A: It certainly is!
B: It couldn't be, and even if it were true, it could never be proved!

Goodwin Sammel quoted by Raymond Smullyan [Smu83]

Electronic computers are mostly used to automate clerical tasks, but it is well known that they can also be programmed to play chess, compose music, and prove mathematical theorems. Since logical, mathematical reasoning is one of the purer forms of human, intellectual thought, the automation of such reasoning by means of electronic computers is a basic and challenging scientific problem. This book is about how computers can be used to construct and check mathematical proofs. This use of computers relies on their symbolic and deductive capabilities and is radically different from their use as fast numerical calculators. The impact of computers as intellectual, rather than clerical, tools is likely to be quite profound. This book tries to demonstrate in a concrete way that the capabilities of computing machines in this regard are quite powerful. It describes proofs of some significant theorems of mathematical logic that were completely checked by means of a computer.

More specifically, this book describes the use of the Boyer–Moore theorem prover in verifying proofs of Gödel's first incompleteness theorem and the Church–Rosser theorem of the lambda calculus. These theorems are among the landmarks in a branch of mathematics called metamathematics. The mechanical verifications of these theorems was carried out as part of my 1986 doctoral dissertation at the University of Texas at Austin. This dissertation addressed the problem of whether automated reasoning technology could be employed to formally prove some substantial piece of mathematics. Gödel's incompleteness theorem was chosen as the natural challenge. It is obvious that this theorem can, in principle, be formally proved. My goal was to determine whether this proof could, in practice, be formalized and checked using a computer program. A secondary goal was to determine the effort involved in such a verification, and to identify the strengths and weaknesses of automated reasoning technology.

The Boyer–Moore theorem prover is a computer program for proving properties of recursive programs. This theorem prover is known for its powerful heuristics for constructing proofs by induction while making clever use of previously proved lemmas. The Boyer–Moore theorem prover did not discover proofs of the incompleteness theorem but merely checked a detailed but fairly high-level proof containing over 2000 definitions and lemmas leading to the main theorems. These definitions and lemmas were constructed through a process of interaction with the theorem prover which was able to automatically prove a large number of nontrivial lemmas. By thus proving a well-chosen sequence of lemmas, the theorem prover is actually used as a *proof checker* rather than a theorem prover.

If we exclude the time spent thinking, planning, and writing about the proof, the verification of the incompleteness theorem occupied about eighteen months of effort with the theorem prover. The verification of the Church–Rosser theorem occupied less than two months of effort. The core result in the latter proof was obtained in a matter of days. Thus though these proofs were successfully verified, the level of effort required was quite considerable. Few mathematicians would be willing to invest this much time and effort in checking their results. The main difficulty in verifying these proofs was the fact that these proofs had to be developed almost entirely from scratch. There was very little background knowledge already available in the theorem prover that could have been utilized for these verifications.

The machine verification of the incompleteness theorem and the Church-Rosser theorem did not contribute much in the way of new and significant knowledge regarding these proofs. Though the mechanical proofs are not directly based on any particular prior informal exposition of these theorems, they cannot be regarded as novel proofs. This is hardly surprising since one secondary goal of the project was to ensure that the verification did not resort to any unconventional short cuts. That is, the intent of the project was to verify a conventional proof of these theorems rather than to prove these theorems by any means possible. Our exposition of the incompleteness theorem deviates from (and improves upon) conventional expositions in several ways. The proof here establishes the incompleteness of a finite set theory so that the Gödel-encoding of syntax is quite perspicuous. Our proof demonstrates that all partial (and not merely total) recursive functions can be represented in this finite set theory. The metatheory employed in these proofs is the highly constructive, quantifier-free theory of Lisp functions employed by the Boyer–Moore theorem prover. It is clear that a considerable amount of metamathematics can be conveniently formalized using this metatheory. In the proof of the incompleteness theorem presented here, Lisp has also been used to characterize the notion of computability. These uses of Lisp have been previously advocated by John McCarthy, the inventor of Lisp.

Proofs in metamathematics are interesting because they can be used to improve the power and sophistication of automated reasoning systems by allowing complex deduction algorithms to be proven sound. For example, the tautology theorem asserts that all tautologies are provable. By using this theorem, a given statement that is a tautology is seen to be provable without explicitly constructing its proof. Thus the mechanization of metamathematical proofs is a crucial step towards more powerful and flexible automated deduction systems.

This book does not contain much discussion of the philosophical consequences of a machine-checked proof of Gödel's incompleteness theorem. The project described here is perhaps of little relevance to the debate on machines and intelligence. My own philosophical views on these matters are not at all deeply considered. Work on automated reasoning does contain an inherent bias towards a formalist conception of the foundations of mathematics since a machine does manipulate symbols and expressions according to formal rules. This is not to suggest that mathematics is discovered or created by formal techniques, but that formalism provides the most rigorous means of refuting conjectured theorems or purported proofs. Gödel's results were indeed a setback to some of the larger ambitions of Hilbert's formalist programme, but they also highlighted and amplified the foundational significance of the formalist view. It is the formalist emphasis on syntax and metatheoretic reasoning that makes it possible to represent mathematics in electronic form and to employ computers to verify proofs.

Several writers have criticized the use of computers in mathematical proofs. Tymoczko [Tym79] for instance argues that the computer proof of the four-color theorem [AH76] calls for a fundamental change in our conception of a mathematical proof. De Millo, Lipton, and Perlis [MLP79] argue that computer programs are not worthy of mathematical correctness proofs since such proofs cannot spark much interest in the social process by which mathematical results are evaluated. They also dismiss the idea that mechanized verification can be used to prove any significant piece of mathematics. Fetzer [Fet88] goes even further in arguing that computer programs are physical and not mathematical entities and thereby incapable of being proved correct. Such philosophical criticisms are definitely relevant to the concerns of this book, and are briefly addressed in the concluding chapter.

Who should read this book? It is mainly directed at mathematicians, philosophers, and computer scientists who are interested in the interaction between logic and computing. The sufficiently patient general reader should be able to grasp the details of the proofs of Gödel's incompleteness theorem and the Church–Rosser theorem, and also gain an appreciation for the status of automated reasoning technology. The exposition assumes a comprehension of the logical notation of first-order logic and some exposure to computing on the part of the reader. The Lisp notation of the Boyer–Moore theorem prover is used to describe the formalizations and proofs of various theorems. Whenever Lisp notation is used, it is backed up by informal English commentary. This notation is quite straightforward and cannot be avoided without doing serious damage to the main point of this book: that it is possible to construct and verify a formal argument without sacrificing cogency.

Chapter 1 presents the historical background in logic and automated reasoning, and contains an overview of the book. The Boyer–Moore logic and the accompanying theorem prover are briefly described.

Chapter 2 defines a formal first-order theory of hereditarily finite sets using the Lisp notation of the Boyer–Moore theorem prover. We use this definition to formally state the incompleteness theorem.

Chapter 3 presents a machine verification of the tautology theorem for classical logic. The tautology theorem was perhaps the first significant theorem of metamathematics. It illustrates many of the basic concepts of metamathematics. In the absence of metatheorems such as the tautology theorem, formal proof construction would be excruciatingly tedious.

Chapter 4 outlines the proof of the representability of the metatheory of a theory within the theory itself. This is done by showing that the metatheory is computable and that computations can be represented by means of proofs in the given theory.

Chapter 5 completes the details of the machine verification of a proof of the statement of Gödel's first incompleteness theorem given in Chapter 2.

Chapter 6 describes the mechanical verification of yet another important metamathematical theorem, the Church–Rosser theorem of the lambda calculus.

The concluding chapter discusses the potential benefits of the machine verification of mathematics, especially metamathematics, and also contains a critique of the Boyer–Moore theorem prover.

The complete scripts for the proofs presented here are part of the standard distribution of the Boyer–Moore theorem prover. These scripts and the theorem prover can be obtained electronically via anonymous FTP from the directory /pub/nqthm/ on the Internet host rascal.ics.utexas.edu. To do this, use FTP to log into rascal.ics.utexas.edu as anonymous using the password guest. Use get /pub/nqthm/README and follow the instructions given in that file.

Acknowledgements

This book is based on my doctoral dissertation entitled *Proof Checking Metamathematics* submitted to the Computer Science Department at the University of Texas at Austin in 1986. All of the work reported here was supervised, stimulated, and influenced by Bob Boyer and J Moore to an extent that cannot be overstated. Without their contributions, this work simply would not exist. Woody Bledsoe inspired me to pursue automated reasoning as a specialization. I frequently drew on Norman Martin's inexhaustible store of knowledge on mathematical logic. The other members of my thesis committee: Allen Emerson, Chris Lengauer, and Matt Kaufmann, also contributed generously to my education, as did Elaine Rich, Jim Browne, Jay Misra, Mani Chandy, the late Lou Rosier, Edsger Dijkstra, Don Good, and several others. During this work, I relied on the support and advice of several friends and colleagues including Bill Bevier, Anne Boyer, Miren Carranza, Shang-Ching Chou, Ernie Cohen, Rich Cohen, David Goldschlag, Warren Hunt, Larry Hines, Pradeep Jain, Myung Kim, Nancy McGough, Don Simon, Werner Uhrig, Mike Vose, and Bill Young. Mike Gordon and Larry Paulson were lively hosts during a very stimulating visit to Cambridge University in 1984. Rob Witty, who was then at SERC(UK), was responsible for arranging this visit. Gian Luigi Bellin, Bob Constable, Gerard Huet, Nicolas de Bruijn, Jon Barwise, Sol Feferman, John McCarthy, Carolyn Talcott, Grigori Mints, and Dana Scott have been important sources of advice and encouragement. Narciso Marti-Oliet and Sam Owre proofread the draft with extreme care and suggested several important revisions. Sam Owre also helped me with some of the nuances of LaTeX and Emacs. Pat Lincoln, Paliath Narendran, John Rushby, and Mandayam Srivas also gave extensive comments and feedback on previous drafts. The books of Stephen Kleene, Joseph Shoenfield, and Paul Cohen provided significant background material for this research. Donald Knuth's TeX and Leslie Lamport's LaTeX programs were used in typesetting this book, and Richard Stallman's

Emacs editor was used to edit the text. Joseph Goguen of Oxford University and David Tranah of Cambridge University Press were instrumental in preparing this book for publication. The editorial skills of Michael Behrend of Cambridge University Press were most helpful. I am of course solely responsible for any remaining errors, omissions, and distortions in the text.

The work reported in this dissertation was supported by NSF Grant DCR8202943 and ONR Grant 81-K-0634. The Science and Engineering Research Council of the UK sponsored my visit to the University of Cambridge. The preparation of this book was generously supported by the SRI International Computer Science Laboratory. I am deeply indebted to John Rushby, Mark Moriconi, and my other colleagues at SRI for maintaining a supportive and stimulating atmosphere for writing and research.

To my parents, my brother Ramesh, and my sister Shoba, I owe the best and most enjoyable part of my education. My wife Anuradha deserves the final word of appreciation for her charm, patience, and understanding.

Menlo Park, California

Chapter 1

Introduction

*Checking mathematical proofs is potentially
one of the most interesting and useful
applications of automatic computers.*
John McCarthy [McC62]

Very few mathematical statements can be judged to be true or false solely by means of direct observation. Some statements, the *axioms*, have to be accepted as true without too much further argument. Other statements, the *theorems*, are believed because they are seen as logical consequences of the axioms by means of a *proof*. Proofs constitute the only effective mechanism for revealing truth in mathematics. We would naturally hope that all provable statements are indeed true. Most of us would also optimistically believe that any true statement has a proof, but such is not the case. Gödel showed that for any reasonably powerful formal system of axioms and inference rules, there are statements that can neither be proved nor disproved on the basis of the axioms and inference rules, and are therefore *undecidable*. Gödel also showed that for such formal systems, there could be no absolute proof that all provable statements were in fact true. In his proof, Gödel described a machine that could check if a given construction constituted a valid proof. It was hoped that one could similarly define a machine to discover the proof itself, but Church and Turing showed that such a machine could not exist. We show in this book that a machine, the Boyer–Moore theorem prover, can be used to check Gödel's proof of the existence of undecidable sentences. Our mechanical verification of Gödel's proof can be seen as an instance of a machine establishing the limitation of mechanism itself, thus showing in concrete terms that machines are not, in this sense, limited.

The notion of a proof as a *logical deduction* of a theorem from the *axioms* was first popularized by Euclid in his *Elements* [Euc56]. For a long time thereafter, there was no rigorous definition of what constituted a valid logical deduction. In the seventeenth century, Leibniz advanced the notion of a universal symbolic logic that he hoped would be applied to mathematical and scientific reasoning, and also to philosophy, law, and politics.[1] In the middle of

---

[1]Leibniz's actual words are memorable [Lei65]:

What must be achieved is in fact this: that every paralogism be recognized as an *error of calculation*, and every *sophism* when expressed in this new kind of notation, appear as a *solecism* or *barbarism*, to be corrected easily by the laws of this philosophical grammar.

Once this is done, then when a controversy arises, disputation will no more be needed between

the nineteenth century, Boole, together with de Morgan, laid out the algebraic laws obeyed by truth-valued propositions and the propositional connectives [Boo54]. In the second half of the nineteenth century, Dedekind developed an axiomatic system for proving properties of numbers, and Cantor developed set theory as a framework for all of mathematics. The first rigorous definition of a valid logical proof was given by Frege in his *Begriffsschrift* [Fre67], where he gave syntactic rules for distinguishing the valid proofs from the invalid ones.[2] He devised a symbolic language, the *predicate calculus*, in which mathematical statements could be unambiguously expressed in terms of variables, functions, predicates, propositional connectives, and universal and existential quantification. In formulations of the predicate calculus, some statements, such as "*A implies A*," are taken to be axioms. Theorems are derived starting from the axioms by repeated application of the rules of inference which prescribe how proofs are to be constructed. The well-known rule of *modus ponens* allows the derivation of a proof of *B* from proofs of *A* and *A implies B*. The derivation of a theorem starting from the axioms and using the rules of inference is termed a *formal proof* since it is defined solely in terms of the *form* or the syntax of the statements involved. The language, axioms, and rules constitute a *formal theory*. Examples of concepts formalized by formal theories include geometry, numbers, and sets.

Once the notion of a proof has been made formal, two important consequences follow. First, a computer can be programmed as an *automated proof checker* to check whether a purported formal proof is correct according to the axioms and inference rules. Second, these formal proofs can themselves be made the objects of mathematical study. This study was given the name *metamathematics* by David Hilbert. Theorems in metamathematics typically analyze the properties of formal theories, their inter-relationships, and the relationship between the form and the provability of statements within specific theories. For example, the tautology theorem asserts that all tautologously true statements of predicate calculus can be formally proved. Similarly, Gödel's first incompleteness theorem asserts of a theory such as formal number theory that it is either inconsistent or contains an undecidable sentence that can neither be proved nor disproved.

This book is about the interplay between automated proof checking and metamathematics. We describe a project aimed at constructing and mechanically verifying several substantial proofs in metamathematics using an automated proof checker known as the Boyer–Moore theorem prover.[3] The proofs thus formally verified include some of the landmarks

---

two philosophers than between two computers. It will suffice that, pen in hand, they sit down to their abacus and (calling in a friend, if they so wish) say to each other: *let us calculate*.

[2]To quote from van Heijenoort's prefatory remarks in *From Frege to Gödel* [vH67]:

This is the first work that Frege wrote in the field of logic, and, although a mere booklet of eighty-eight pages, it is perhaps the most important single work ever written in logic. Its fundamental contributions, among lesser points, are the truth-functional propositional calculus, the analysis of the proposition into function and argument(s) instead of subject and predicate, the theory of quantification, a system of logic in which derivations are carried out exclusively according to the form of the expressions, and a logical definition of the notion of a mathematical sequence. Any one of these achievements would suffice to secure the book a permanent place in the logician's library.

[3]It could be argued that the mechanically verified proofs described here are not conventional formal proofs

of metamathematics: the tautology theorem, Gödel's first incompleteness theorem, and the Church–Rosser theorem of the lambda calculus. We thus demonstrate that the technology of automated proof checking is sufficiently well-developed that it is possible to verify substantial proofs in metamathematics. If complex and substantial mathematical arguments can be mechanically verified, then it is conceivable that automated proof checking technology will eventually be used as a reasoning aid and even employed as part of the refereeing process for journal publications. The technology is not yet ripe enough for such a task but there are no obvious insurmountable obstacles. Mechanically verified proofs in metamathematics themselves have significant relevance for automated proof checking. Theorems in metamathematics make it possible to obtain more sophisticated but sound inference procedures from simpler ones. The ability to add such inference rules is essential if high-level mathematical arguments are to be verified with a reasonable amount of effort.[4]

The remainder of this introduction presents the background material in metamathematics and automated reasoning and provides an overview of the chapters to follow. We also give a brief introduction to the Boyer–Moore theorem prover and its logic.

1.1   Background

The relevant literature on logic, metamathematics, and automated reasoning is obviously voluminous and only a few significant sources are enumerated here. One of the more readable outlines of Gödel's incompleteness theorem is Gödel's original paper [Göd67b] entitled "On formally undecidable propositions in Principia mathematica and related systems I."[5] This paper has been widely reprinted and an English translation [Göd92] has recently been published by Dover Publications as a book in 1992. Two good sources for Gödel's paper and related material are:

- *The Undecidable*, edited by Davis [Dav65], and

- *From Frege to Gödel: A Sourcebook in Mathematical Logic, 1879–1931*, edited by van Heijenoort [vH67].

Gödel's paper also appears in *Kurt Gödel: Complete Works Volume I* edited by Feferman et al. [FJWDK+86]. Topics related to the incompleteness theorem are discussed by Smorynski [Smo78] and by Smullyan [Smu92] who has also given entertaining explanations of Gödel's theorems in several popular books.

Kleene's *Introduction to Metamathematics* [Kle52], Shoenfield's *Mathematical Logic* [Sho67], and Cohen's *Set Theory and the Continuum Hypothesis* [Coh66], provided some of the primary source material for the present project.

---

in any well-known formal proof system since the Boyer–Moore theorem prover employs a number of complex decision procedures and heuristics. However, these proofs are, both in principle and in practice, formalizable.

[4]Frege foresaw this possibility when he wrote [Fre67]:

> ...when the *foundations* for such an ideography are laid, the primitive components must be taken as simple as possible, if perspicuity and order are to be created. This does not preclude the possibility that *later* certain transitions from several judgments to a new one, transitions that this one mode of inference would not allow us to carry out except mediately, will be abbreviated in immediate ones. In fact this would be advisable in case of eventual application.

[5]Since Gödel's results were quickly accepted, the planned second part of the paper was never written.

The main references on the Boyer–Moore theorem prover are the books *A Computational Logic* [BM79] and *A Computational Logic Handbook* [BM88], by Boyer and Moore. The book *Symbolic Logic and Mechanical Theorem Proving* by Chang and Lee [CL73] is a good reference for material on resolution theorem proving.

Other general texts on mathematical logic include *A Mathematical Introduction to Logic* by Enderton [End72], and *Computability and Logic* by Boolos and Jeffrey [BJ89].

Books on the lambda calculus include Church's *The Calculi of Lambda-Conversion* [Chu41], Barendregt's *The Lambda Calculus* [Bar78a], and *Introduction to Combinators and λ-Calculus* by Hindley and Seldin [HS86].

The rest of this section informally presents the relevant background on language, logic, metamathematics, and automated reasoning.

1.1.1 The Paradoxes

Dedekind, Cantor, Peano, and Frege were seeking a uniform foundational framework for mathematics in terms of a language and a deductive system. The difficulty of this task can be illustrated by means of several well-known paradoxes. Many of these exploit the confusion between language, metalanguage, and semantics. The Liar paradox [Mar84] has a Cretan asserting, "Cretans always lie." If indeed Cretans always lie, then the above statement is true. Then, however, since this statement is true and spoken by a Cretan, it cannot be the case that Cretans always lie. So if the sentence is true, then it is also false. On the other hand, if there is exactly one Cretan whose only utterance is the above statement, then if this statement is false, it is also true. The Liar paradox is a semantical paradox since it employs notions of meaning, truth, and falsity in its statement. The paradox suggests that these notions are not accurately definable for a language within the language itself [Tar83].

Berry's paradox defines a number as "the least natural number not describable in fewer than ninety letters from the English alphabet." Since there are only finitely many numbers describable in fewer than ninety letters from the English alphabet, a least such number must exist. However, if such a number does exist, then we have in fact succeeded in describing it in fewer than ninety letters by the phrase in quotes. Berry's paradox is another semantical paradox.

Russell's paradox evokes the idea of a library catalog that lists all those library catalogs that do not list themselves. If this catalog does not list itself, then it would be incomplete and hence would not have listed *all* of those catalogs that do not list themselves. On the other hand, if it does list itself, then it is erroneous since it has then listed a catalog that lists itself. Russell's paradox can actually be put into a mathematical form. It does not employ any obviously semantical notions, and it yields a contradiction in Frege's theory of arithmetic [GB80, Rus67].

Thus, paradoxes do appear in seemingly natural formalizations of mathematical reasoning and they lead, as shown above, to contradictions. The challenge for a foundation of mathematics is to formalize all of mathematics while avoiding the contradictions resulting from such paradoxes. The constructions of undecidable sentences in proofs of Gödel's incompleteness theorem are also inspired by the paradoxes.

1.1.2 Foundations of Mathematics

If mathematics relies on logical deduction rather than direct observation as a means of grasping the truth, then, as the above paradoxes demonstrate, it is crucial to identify the principles of correct mathematical reasoning. Euclid [Euc56] identified such a collection of "self-evident" postulates for proving theorems in plane geometry, but used informal mathematical reasoning to construct proofs. The discovery of non-Euclidean geometry in the early nineteenth century cast doubt on the self-evidence of Euclid's postulates. This discovery and the general increase in mathematical rigor led several mathematicians to carefully identify certain correct reasoning principles and to ensure that these were in some sense minimal. The hope was that such a systematization of mathematical reasoning would also make it possible to find the errors in a mathematical proof in a systematic manner. Dedekind [Ded63] carried out such a development for number theory where he carefully developed axioms for the natural numbers and reduced the theory of rational and real numbers to a combination of set theory and number theory. Peano [Pea67] gave number-theoretic axioms similar to those of Dedekind and employed an elegant logical notation that has since become standard in mathematics. Cantor [Can55] went even further and reduced a great deal of mathematics to a set theory that contained transfinite ordinal and cardinal numbers.

Logicism. Towards the end of the nineteenth century, the question arose as to whether all mathematical truths could be deduced purely from simple, self-evident logical principles. This was Frege's aim when he first set out the predicate calculus in his *Begriffsschrift* [Fre67] in 1879, and later attempted to develop arithmetic from purely logical principles. Russell found an inconsistency in Frege's axiomatization of arithmetic. Whitehead and Russell stuck to the logicist line and attempted to develop a purely logical theory of mathematics in the *Principia Mathematica* [WR25]. They developed a somewhat baroque theory of types and showed that a significant amount of mathematics could be rigorously derived from their axioms. It would be difficult to argue that the system of *Principia Mathematica* employed only logical principles. There have been attempts to revive logicism but none that are very convincing.

Intuitionism. In the late nineteenth and early twentieth century, Kronecker, then Poincaré, and eventually Brouwer, became heavily skeptical about the purity of the methods of proof employed in large parts of mathematics. The crux of the attack was with the use of infinite entities such as the set of natural numbers. Mathematics is replete with proofs where the existence of a mathematical object satisfying a property is demonstrated by showing that its non-existence yields a contradiction.[6] This, the critics argued, is fine when the domain is finite since all possible candidates can be examined in order to find one satisfying the required property. No such exhaustive search is possible with infinite domains and hence the argument by contradiction does not yield a constructive demonstration of existence.

Brouwer's critique led him to formulate intuitionism as a pure approach to mathematics that only used constructive methods and treated infinite structures as potential rather

---

[6]The classic example is a proof that there exist irrational numbers x and y such that x^y is rational. The argument is that either $\sqrt{2}^{\sqrt{2}}$ is rational, in which case x and y can both be taken as $\sqrt{2}$, or we pick x to be $\sqrt{2}^{\sqrt{2}}$ and y to be $\sqrt{2}$ so that x^y is just $\sqrt{2}^{\sqrt{2}^2}$ which simplifies to 2.

than completed infinities. The intuitionistic approach to mathematics has been studied and further developed by a number of mathematicians including Kolmogorov, Heyting, Bishop, Markov, Shanin, and many others [TvD88]. The use of constructive proof techniques has also had a significant influence on computer science, particularly through proof checking tools such as Nuprl [Con86]. Intuitionism does place serious and pervasive limitations on the methods of mathematics and has not yet had much of an impact on mathematical practice.

Formalism. Hilbert reacted to the intuitionistic critique by initiating a research programme to demonstrate that conventional mathematical methods were in fact valid. He wished to prove this using rigorous proof methods of mathematics that were acceptable beyond any shadow of doubt. His claim was that the use of infinities in mathematics was a convenient idealization and that any concrete conclusions about numbers that were drawn from such idealizations could be shown to be correct. Hilbert hoped that such a metamathematical study would show that mathematics was free of contradiction and hence *consistent*. In developing his metamathematics, he identified mathematics with a formal game played on paper with symbols and syntactic rules for forming statements and proofs. Hilbert also posed several other important problems in metamathematics including the question of whether a machine could correctly identify whether a given formal statement was a theorem in a given formal theory of mathematics.

Metamathematics has had a great many successes but Hilbert's original goal was not fulfilled. His own technique of formalizing and arithmetizing the syntax of mathematics led to Gödel's discovery that any reasonable formal theory contained sentences that could not be proved or disproved. As a consequence of this, Gödel showed that any consistency proof of a sufficiently powerful formal theory for mathematics would have to rely on methods that were stronger and more open to doubt than those permitted by the theory. Church [Chu36] and Turing [Tur65] showed that no machine could recognize the theorems in the predicate calculus. Since most formal mathematical theories build on the predicate calculus, the prospect of mechanical *decision procedures* for these theories is not great. There are interesting decidable theories such as Presburger arithmetic which is a form of Peano arithmetic where only addition is defined but not multiplication.

1.1.3   Language and Metalanguage

The quest for a foundation for mathematics begins with the search for a precise and unambiguous notation for expressing mathematics. Modern mathematical notation owes a good part of its precision and polish to the work of logicians such as Frege and Peano. A mathematical language provides a *syntax* for expressing statements about a particular conceptual domain. A *semantics* for a language characterizes those statements that are true of the conceptual domain. The language of *informal* mathematics consists of *ad hoc* notation and everyday language embellished with colloquial mathematical usages. The notation of logic underlies the formalized syntax of mathematics. For example, the axiom of *extensionality* in set theory would be written informally as, "Two sets are equal if they contain exactly the same elements." In formal notation, the same statement would be

$$(\forall x, y. \ (\forall z. \ z \in x \iff z \in y) \supset x = y).$$

Such a language can be given a precise grammar so that it is possible to automatically check an expression for syntactic well-formedness. We can characterize expressions according to their syntactic properties as formulas, atomic formulas, quantified formulas, sentences, and so on. It is possible then to study the connection between the syntactic and semantic properties of expressions. The language in which the syntax and semantics of a formalized language is discussed is called the *metalanguage*. The formal language itself is called the *object language*. Informal mathematical discourse often employs metalinguistic assertions such as, "There are exactly two free variables, '*x*' and '*y*', in the statement of extensionality above." Thus the metalanguage used in mathematical discourse is informal natural language such as English or German.

The work described in the present monograph employs a formal object language, namely that of first-order logic. In contrast to most logic textbooks where the metalanguage is informal, we employ a formal metalanguage that is based on the programming language pure Lisp (see Section 1.1.8).

1.1.4 Logic

Given a *language* for expressing statements about a particular conceptual domain, and a *semantics* for identifying the true statements, a *logic* is a system of *axioms* and *rules of inference* for constructing proofs of statements. Examples of logics include classical and intuitionistic propositional logics, temporal and modal logics, and first-order logic. A *theory* such as number theory or set theory can be formalized within the framework of a logic such as first-order logic by providing additional axioms. A statement is *provable* in a logic if it has a proof constructed from the axioms using the rules of inference. The provable statements are the *theorems* of the logic. A statement is *disprovable* if its negation is provable. A logic is *sound* if all provable statements are semantically true. It is *complete* if all of the semantically true statements are provable. The logic is *consistent* if no statement and its negation are both provable. A logic is *decidable* if there is an effective algorithm to determine whether a given statement is a theorem.

Propositional logic, for example, is about propositions, which as the dictionary indicates are expressions "in language or signs of something that can be believed, doubted, or denied or is either true or false." There are many ways that propositional logic can be presented; the presentation below follows Shoenfield [Sho67]. *Formulas* in the language consist of:

- the *propositional atoms* such as p, q, and r,

- *negations*: $(\neg A)$ (read as "not A"), where A is a formula, and

- *disjunctions*: $(A \vee B)$ (read as "A or B"), where A and B are themselves formulas.

Except for the quantified formulas to be introduced below, the surrounding parentheses will be dropped when they contribute nothing to the readability of the formula. Note that negation binds the tightest, and that conjunction and disjunction bind more tightly than implication and equivalence. The other propositional connectives can easily be defined in terms of negation and disjunction:

- *Implication*: $A \supset B \equiv \neg A \vee B$ (read as "A implies B").

A	$\neg A$
true	**false**
false	**true**

A	B	$A \vee B$
true	**true**	**true**
true	**false**	**true**
false	**true**	**true**
false	**false**	**false**

Figure 1.1: Truth Tables for Negation and Disjunction

$$\frac{}{A \vee \neg A} \; \textit{Axiom}$$

$$\frac{B}{A \vee B} \; \textit{Weakening}$$

$$\frac{A \vee A}{A} \; \textit{Contraction}$$

$$\frac{A \vee (B \vee C)}{(A \vee B) \vee C} \; \textit{Associativity}$$

$$\frac{A \vee B \quad \neg A \vee C}{B \vee C} \; \textit{Cut}$$

Figure 1.2: Proof Rules for Propositional Logic

- *Conjunction*: $A \wedge B \equiv \neg(\neg A \vee \neg B)$ (read as "A and B").

- *Equivalence*: $(A \Longleftrightarrow B) \equiv (A \supset B) \wedge (B \supset A)$ (read as "A equivales B" or "A if and only if B").

The classical semantics for propositional formulas is given by assigning *truth values*, **true** or **false**, to the propositional atoms and evaluating the statements for each such assignment using the *truth-table interpretation* of negation and disjunction. These truth tables are displayed in Figure 1.1. The statements of propositional logic are just its formulas. The true statements are those that evaluate to **true** under any assignment of truth values to the propositional atoms. A typical proof rule of the propositional logic, the *law of the excluded middle*, asserts that any formula of the form $A \vee \neg A$ is an *axiom*. Such a proof rule is an axiom scheme since it asserts $A \vee \neg A$ to be an axiom for any formula A. Any formula of the form $A \vee \neg A$ always evaluates to **true** regardless of whether A is assigned **true** or **false**.

The excluded middle axiom and the other proof rules of propositional logic are shown in Figure 1.2. For example, the proof rule of *weakening* would yield a proof of $(A \vee B)$ from a proof of B. Propositional logic can easily be shown to be sound relative to the truth-table interpretation, and also to be consistent and complete.[7]

Furthermore, propositional logic is *decidable*: there is a straightforward algorithm to decide if a given statement is a theorem by simply checking if every truth-table evaluation of the statement is **true**. The consistency, completeness, and decidability of propositional logic

---

[7]Soundness typically implies consistency. A sound theory is consistent if the semantics does not judge some statement and its negation to both be true. A theory could be consistent in not proving some statement and its negation but could be unsound relative to the intended semantics.

constitute the earliest significant contributions to metamathematics and were first proved in Post's doctoral dissertation [Pos21], and independently by Bernays [Ber26].

First-order logic or the predicate calculus is a refinement of propositional logic where the truth values of propositional atoms are allowed to vary depending on the values assigned to the *individual variables*. The syntax of first-order logic contains individual variables, function symbols (e.g., the addition and multiplication operations of arithmetic), predicate symbols (e.g., the symbols for the ordering relations $<$ and $\leq$ on numbers), and quantifiers. First-order *terms* are either

- individual variables denoted by x, y, and z, or

- of the form $f(t_1, \ldots, t_n)$, where f is an n-ary *function symbol* and the t_i are terms.

An *atomic formula* has the form $p(t_1, \ldots, t_n)$, where p is an n-ary predicate symbol and the t_i are terms. The arities of predicates and function symbols can be zero. The *formulas* include

- atomic formulas,

- the *negation* $\neg A$ of any formula A,

- the *disjunction* $A \vee B$ of any two formulas A and B, and

- the *existential quantification* $(\exists x.\ A)$ (read as "for some x, A") of any formula A with respect to a variable x.

The form $(\forall x.\ A)$ (read as "forall x, A) can be defined as $\neg(\exists x.\ \neg A)$. Unlike propositional logic, where the meaning of an atomic proposition in a given interpretation is either **true** or **false**, the meaning of $p(t_1, \ldots, t_n)$ varies according to the meanings of the t_i. A *first-order language* is a collection of function and predicate symbols obeying the above syntactic rules.

Any expression (term or formula) is a *subexpression* of itself; the subexpressions of $f(t_1, \ldots, t_n)$ and $p(t_1, \ldots, t_n)$ also include the subexpressions of the t_i, the subexpressions of $\neg A$ and $(\exists x.\ A)$ include those of A, and the subexpressions of $A \vee B$ include those of A and B. A formula A is a *subformula* of a another formula B if A is a subexpression of B. A variable x is said to *occur* in a term or formula in which it occurs as a subexpression. Any occurrence of x in A is *bound* in $(\exists x.\ A)$. An occurrence of a variable is *free* in a formula if it is not bound in any subformula of the given formula. Any variable that has a free occurrence in a formula is a free variable of the formula. A *statement* or a *sentence* is a formula with no free variables.[8] The notations $[r/x]t$ and $[r/x]A$ refer to the results of substituting the term r for all free occurrences of the variable x in the term t and the formula A, respectively. First-order logic *with equality* contains a special 2-ary predicate symbol for equality that is used in the infix form $a = b$, where a and b are terms.

A *model* for a first-order language assigns meanings to the function and predicate symbols relative to a nonempty domain D. A model $\mathcal{M}$ associates each n-ary function symbol in the language to a mapping from D^n to D, and each n-ary predicate symbol in the language to

---

[8] Any formula can be seen as a statement where the free variables in the formula are implicitly universally quantified at the outermost level. So if A is a formula in a proof and x is the only free variable in A, then the statement A is read implicitly as $(\forall x.\ A)$.

$$\begin{aligned}
\mathcal{M}[\![x]\!]\rho &= \rho(x) \\
\mathcal{M}[\![f(t_1,\ldots,t_n)]\!]\rho &= \mathcal{M}(f)(\mathcal{M}[\![t_1]\!]\rho,\ldots,\mathcal{M}[\![t_n]\!]\rho) \\
\mathcal{M}[\![a = b]\!]\rho &= \begin{cases} \textbf{true}, & \text{if } \mathcal{M}[\![a]\!]\rho = \mathcal{M}[\![b]\!]\rho \\ \textbf{false}, & \text{otherwise} \end{cases} \\
\mathcal{M}[\![p(t_1,\ldots,t_n)]\!]\rho &= \mathcal{M}(p)(\mathcal{M}[\![t_1]\!]\rho,\ldots,\mathcal{M}[\![t_n]\!]\rho) \\
\mathcal{M}[\![\neg A]\!]\rho &= \begin{cases} \textbf{false}, & \text{if } \mathcal{M}[\![A]\!]\rho = \textbf{true} \\ \textbf{true}, & \text{otherwise} \end{cases} \\
\mathcal{M}[\![A \vee B]\!]\rho &= \begin{cases} \textbf{true}, & \text{if } \mathcal{M}[\![A]\!]\rho = \textbf{true} \\ \textbf{true}, & \text{if } \mathcal{M}[\![B]\!]\rho = \textbf{true} \\ \textbf{false}, & \text{otherwise} \end{cases} \\
\mathcal{M}[\![(\exists x.\ A)]\!]\rho &= \begin{cases} \textbf{true}, & \text{if } \mathcal{M}[\![A]\!][d/x]\rho = \textbf{true}, \text{ for some } d \text{ in } D \\ \textbf{false}, & \text{otherwise} \end{cases}
\end{aligned}$$

Figure 1.3: Semantics of First-Order Logic

$$\overline{[a/x]A \supset (\exists x.\ A)}\ {}^{Substitution}$$

$$\overline{a = a}\ {}^{Identity}$$

$$\overline{a_1 = b_1 \supset (\ldots \supset (a_n = b_n \supset f(a_1,\ldots,a_n) = f(b_1,\ldots,b_n)))}\ {}^{Equality_1}$$

$$\overline{a_1 = b_1 \supset (\ldots \supset (a_n = b_n \supset p(a_1,\ldots,a_n) \supset p(b_1,\ldots,b_n)))}\ {}^{Equality_2}$$

$$\{x \text{ is not free in } B\}\frac{A \supset B}{(\exists x.\ A) \supset B}\ {}^{\exists\text{-}Introduction}$$

Figure 1.4: (Non-Propositional) Proof Rules for First-Order Logic

a mapping from D^n to the truth values $\{\textbf{true}, \textbf{false}\}$. A model by itself is not enough to determine the meanings of expressions since we also need to fix the interpretation of any free variables in the expression. An *assignment* ρ assigns values in D to the individual variables of the language. A formula A is true in a model if it is true with respect to any assignment ρ of values in D to the free variables in A. If ρ is an assignment, then $[d/x]\rho$ is the assignment that assigns d to x and $\rho(y)$ to any other variable y. The meaning of a formula in a model $\mathcal{M}$ under an assignment ρ is shown in Figure 1.3. The formula $(\exists x.\ A)$ is true in the model under assignment ρ if the formula A is true in the model under the assignment $[d/x]\rho$, for some value d in D. The truth value of a formula with free variables depends on the choice of the assignment, but a statement can only be either true or false in a model. A statement is semantically true if it is true in all models.

In addition to the proof rules of propositional logic (in Figure 1.2), first-order logic contains rules pertaining to equality and quantification. These rules are shown in Figure 1.4. It is easy to show that first-order logic is sound by checking that the axioms are satisfied in any nonempty model, and that for each rule of inference, the conclusions are true in the model if the premises are true in the model. Gödel [Göd67a] showed that first-order logic is also complete: any first-order logic statement that is true in all models is also provable from the rules and axioms of first-order logic. Church [Chu36] and Turing [Tur65] independently demonstrated that first-order logic is algorithmically undecidable. First-order logic has a number of other interesting properties [Bar78b].

1.1.5 Peano Arithmetic

First-order logic formalizes some general principles of reasoning independent of any particular conceptual domain. In order to formalize particular concepts such as numbers or sets, it is necessary to add certain *nonlogical* axioms to first-order logic. The formal theory of Peano arithmetic (actually due to Dedekind [Ded63]) introduces the nonlogical axioms needed to prove theorems in number theory.[9] The language of Peano arithmetic contains the constant symbol 0 denoting the number zero, and the function symbol S used to construct the successor $S(n)$ of a numeral n. The axioms, $0 \neq S(x)$ and $S(x) = S(y) \supset x = y$, essentially are tools for demonstrating the inequality of two numerals. Addition is defined by the two axioms, $0 + y = y$ and $S(x) + y = S(x + y)$. Multiplication is similarly defined by the two axioms, $0 \times y = 0$ and $S(x) \times y = y + (x \times y)$. Induction is given by an axiom *scheme*, a family of axioms (one for each formula A) of the form:

$$([0/x]A \wedge (\forall x.\ (A \supset [S(x)/x]A))) \supset (\forall x.\ A),$$

where A is any formula in the language of Peano arithmetic.

Peano arithmetic (PA) would be complete if it could be used to prove all the true statements of the conceptual domain of numbers. However, Gödel's first incompleteness theorem asserts that PA cannot be both complete and consistent. This theorem remains true of any reasonable extension of PA, and is also applicable to a wide class of formal theories that are rich enough to express a limited amount of arithmetic. Gödel's second incompleteness theorem shows that though the metamathematical statement of the consistency of PA can be formalized in PA, it cannot be proved unless PA is inconsistent (in which case all statements are provable). Gentzen [Gen69] was in fact able to demonstrate the consistency of PA by using metamathematical reasoning principles that cannot be formalized within PA. Gödel's argument can easily be adapted to show that Peano arithmetic is also undecidable. If either addition or multiplication is dropped from Peano arithmetic, the resulting theory becomes decidable. More dramatically, Paris and Harrington [PH78] have demonstrated the existence of a natural combinatorial mathematical statement, a variant of the celebrated Ramsey theorem, that can be proved in set theory but not in Peano arithmetic.

1.1.6 Set Theory

The intent of Cantor's set theory [Can55] was to provide a conceptual framework for all of mathematics. Set theory has actually been fairly succcessful from this point of view; the

---

[9]Both Dedekind [Ded63] and Peano [Pea67] used some set theory in formulating their number-theoretic axioms so that they did not need definitions of addition and multiplication. The version of Peano arithmetic here avoids the use of sets but builds in the definitions of addition and multiplication.

$$\overline{x = y \iff (\forall z.\ z \in x \iff z \in y)}^{\ Extensionality}$$

$$\overline{\neg(x \in \emptyset)}^{\ NullSet}$$

$$\overline{z \in \{x, y\} \iff z = x \lor z = y}^{\ Pairing}$$

$$\overline{z \in x \cup y \iff z \in x \lor z \in y}^{\ Union}$$

$$\overline{Trans(x) \iff (\forall y, z.\ y \in x \land z \in y \supset z \in x)}^{\ Transitivity}$$

$$\overline{Linear(x) \iff (\forall y, z.\ y \in x \land z \in x \supset y = z \lor y \in z \lor z \in y)}^{\ Linear}$$

$$\overline{I(x) \iff Trans(x) \land Linear(x)}^{\ Number}$$

$$\overline{([\emptyset/x]A \land (\forall x.\ ((I(x) \land A) \supset [S(x)/x]A))) \supset (\forall x.\ I(x) \supset A)}^{\ Induction}$$

Figure 1.5: Axioms of Z2

language and notation of set theory are widely used in mathematics. There are two first-order formalizations of set theory. One is due to Zermelo, Skolem, and Fraenkel, and is called ZF. The other is due to Bernays, von Neumann, and Gödel (called BVG) which is a 2-sorted theory of sets and *proper classes* which can have elements that are sets but cannot themselves be elements of other sets or classes.

There are two basic relations in ZF set theory: equality ($=$) and membership ($\in$). The extensionality axiom of ZF asserts that sets are determined solely by the collection of their elements so that two sets with identical elements are equal. One basic way to construct sets is to use the *comprehension* axiom scheme to define the set $\{x \mid \phi(x)\}$ of all those elements that satisfy a given property ϕ. This naive form of comprehension is unsound in that it falls prey to Russell's paradox: the set R defined as $\{x \mid x \notin x\}$ leads to both $R \in R$ and $R \notin R$ being true. Zermelo's version of the comprehension scheme only permits sets to be defined by comprehension by collecting those elements of some pre-existing set that satisfy a given property. Apart from comprehension, ZF allows sets to be constructed as the null set, the power set (i.e., the set of all subsets) of a given set, the image of a set under a rule given by a formula (the replacement axiom scheme), and as the set ω of all the natural numbers represented in set-theoretic form. The regularity or foundation axiom ensures that all membership chains are well-founded.

In this book, we restrict our attention to a set theory formalizing a notion of sets that are finite and all of whose elements, elements of elements, etc., are also finite. Such sets are said to be *hereditarily finite*. This set theory, called Z2, is due to Cohen [Coh66] and its axioms are displayed in Figure 1.5. It contains axioms for the operations of forming the null set: $\emptyset$, the (unordered) pair of two sets: $\{a, b\}$, and the union of two sets: $a \cup b$. A singleton set $\{a\}$ is formed as the pair $\{a, a\}$. The Weiner–Kuratowski definition of ordered pairs $\langle a, b \rangle$ is given by $\{\{a, b\}, \{a\}\}$. The first and second projections of the ordered pair $\langle a, b \rangle$ can easily be defined to return a and b, respectively. The von Neumann definition for finite ordinals (natural numbers) can be given as $O(0) = \emptyset$ and $O(n + 1) = \{O(n)\} \cup O(n)$.

Apart from the extensionality axiom, there are three axioms of transitivity, linearity, and induction for finite ordinals. As formal theories of arithmetic, Z2 and PA have the same expressive power; there is a translation between the purely arithmetic theorems of Z2 and PA. Therefore Z2 is also incomplete and undecidable. The main result of this book is a mechanical verification of the incompleteness of Z2. Z2 is more convenient than PA for this purpose since Z2 contains an ordered pairing operation that is tedious to encode purely in terms of numbers.

1.1.7 Computability

One significant outcome of Gödel's proof was the formalization of the notion of *effective computability* in terms of *recursive functions*. Gödel's proof of the incompleteness theorems employed the notion of *primitive recursive functions*, the class of functions that can be defined in one of the following forms:

- constant functions of the form $f(x_1 \ldots x_n) = c$, for some constant c

- projection functions of the form $f(x_1 \ldots x_n) = x_i$, for some i, $0 < i \le n$

- the successor function of the form $S(x) = x + 1$

- composition of primitive recursive functions $g, h_1, \ldots, h_m$, of the form:

$$f(x_1 \ldots x_n) = g(h_1(x_1 \ldots x_n), \ldots, h_m(x_1 \ldots x_n))$$

- recursive definition in terms of primitive recursive functions g and h, of the form:

$$
\begin{aligned}
f(0, x_1 \ldots x_n) &= g(x_1 \ldots x_n) \\
f(S(y), x_1 \ldots x_n) &= h(y, f(y, x_1 \ldots x_n), x_1 \ldots x_n)
\end{aligned}
$$

For example, the definitions of the addition and multiplication operations of Peano arithmetic fall within the class of primitive recursive functions. There are examples of computable functions that are not primitive recursively definable. The best known of these is Ackermann's function which exhibits a lexicographic recursion in both its arguments. Péter's [Pét67] version of Ackermann's function can be defined as:

$$
\begin{aligned}
ack(0, n) &= S(n) \\
ack(S(m), 0) &= ack(m, 1) \\
ack(S(m), S(n)) &= ack(m, ack(S(m), n))
\end{aligned}
$$

Herbrand and Gödel defined the class of *general recursive functions* to extend the primitive recursive functions to include functions such as Ackermann's function above. This formulation as presented by Kleene [Kle52] describes arithmetic functions by their set of defining equations as is the case with the definition of Ackermann's function shown above.

Several other formalizations of computability were soon proposed but were all found to be equivalent. For example, the set of primitive recursive functions augmented with a minimization operator $\mu x.f(n,x) = 0$ returning the least value of x such that $f(n,x) = 0$ (provided such an x exists) for a given primitive recursive function f, suffices to define all general recursive functions. Another formulation given in terms of the computations of a *Turing machine* has played a significant role in theoretical computing [Tur65].[10] Turing was able to show that no Turing machine could compute whether a given Turing machine (description) would halt when executed on a given input.

Computability theory is now a substantial branch of mathematics that has enormously influenced computer science. Church's lambda calculus [Chu41] provides yet another definition of computability and it has played a key role in the study of programming languages. The programming language pure Lisp is used here as a characterization of computable functions. Lisp is closely related to the Herbrand-Gödel-Kleene formulation of recursive functions. We are using pure Lisp both as our formal metalanguage and in order to conveniently characterize *partial recursive functions* which are functions that are computable on some or all of their domain values and undefined on any remaining domain values.

1.1.8   Lisp

Lisp (LISt Processing) is a general-purpose programming language created by McCarthy [McC60]. As a programming language, Lisp contains useful but messy destructive operations whose mathematical semantics can be quite complicated [Mas86]. McCarthy [McC63] was able to isolate a mathematically elegant core of the language, pure Lisp, whose main features were:

- the use of *symbolic expressions* as data,

- functions as the primary program unit,

- an if-then-else conditional as the primary construct for controlling the evaluation of expressions, and

- recursion as the primary mechanism for describing unbounded classes of computations.

McCarthy [McC63] also developed an elegant mathematical theory for reasoning about properties of pure Lisp functions. As mentioned earlier, pure Lisp is closely related to the equational computing formalism of Herbrand and Gödel.

Pure Lisp forms the basis for the principles of definition and reasoning used by the Boyer–Moore theorem prover. This latter system provides our formal metalanguage. We also use pure Lisp as a characterization of the class of partial recursive functions in the proof of Gödel's incompleteness theorem. We use the typewriter font for Lisp expressions. Lisp expressions are not case-sensitive: foo, FOO, Foo, etc. are all identical. We follow the convention

---

[10]Alan Turing was one of the first to build a working computer, and to demonstrate that programs could be proved to meet their specifications [MJ84]. Turing was also the first to envisage the possibility that computers could be programmed to perform tasks which require intelligence, such as proving theorems. Another logician, John von Neumann [vN61], was also an important pioneer in the design of early computers and in *program proving*.

of keeping the Lisp expressions in upper case and use lower case to indicate Lisp metavariables to range over Lisp function symbols or expressions. Unlike conventional mathematics where a function application is usually written as $fn(a_1, \ldots, a_n)$, Lisp uses a convention that such a function application be written as (fn $a_1 \ldots a_n$), where fn is a function name, and the a_i are themselves Lisp expressions.

The pure Lisp fragment of Lisp is a programming language for defining operations over the domain of *s-expressions* (symbolic expressions) which includes the numerals representing natural numbers (e.g., 0, 1, 2, ...), the booleans T (for "true") and F (for "false"), the atom NIL,[11] and any ordered pair (s . t) of smaller s-expressions s and t. A list such as (1, 2, 3) is represented in Lisp as (1 . (2 . (3 . NIL))) (which can be further abbreviated as (1 2 3)). Note that (1 . 2) is not the same s-expression as (1 2) since the latter is an abbreviation for (1 . (2 . NIL)).

There are a handful of primitive operations including some basic arithmetic and list operations such as:

- (CONS X Y): constructs the ordered pair whose first element is X and second element is Y.

- (CAR X): returns the first element of X if X is an ordered pair, and 0 otherwise.

- (CDR X): returns the second element of X if X is an ordered pair, and 0 otherwise.

- (LISTP X): returns T if X is an ordered pair, and F otherwise.

- (ADD1 X): adds one to the value of X.

- (SUB1 X): subtracts one from the value of X; returns 0 if X is not a positive number.

- (NUMBERP X): returns T if X is a number, and F otherwise.

- (ZEROP X): returns F if X is a positive number, and T otherwise.

- (EQUAL X Y): returns T if X and Y are identical s-expressions, and F otherwise.

For example, (CONS 1 2) constructs the ordered pair (1 . 2), and (CAR (CONS 1 2)) returns the first component 1, whereas (CDR (CONS 1 2)) returns the second component 2, of the ordered pair.

The conditional expression (IF a b c) returns the value of the expression c if a evaluates to F, and otherwise it returns the value of b. Functions are defined by means of equations using conditional expressions, recursion, and the primitive operations. The function that takes the maximum of two elements is defined as:

    (MAX X Y) = (IF (LESSP X Y) Y X)

where (LESSP X Y) returns T if X is numerically less than Y, and F otherwise. We refer to Lisp functions such as LESSP that only return T or F as values, as predicates.

A wide variety of numeric and list-processing operations can be defined as Lisp functions. The operation of appending two lists is defined recursively as

---

[11] In conventional versions of Lisp, F and NIL are identified, but we follow the Boyer–Moore prover in treating these as distinct.

```
(APPEND X Y) =
    (IF (LISTP X)
        (CONS (CAR X)(APPEND (CDR X) Y))
      Y)
```

The computation of (APPEND (CONS 1 (CONS 2 NIL))(CONS 3 NIL)) can roughly be described as follows ($\Longrightarrow$ denotes 'evaluates to'):

```
        (APPEND (CONS 1 (CONS 2 NIL))(CONS 3 NIL))
    =   (CONS 1 (APPEND (CONS 2 NIL)(CONS 3 NIL)))
    =   (CONS 1 (CONS 2 (APPEND NIL (CONS 3 NIL))))
    =   (CONS 1 (CONS 2 (CONS 3 NIL)))
    ⟹   (1 . (2 . (3 . NIL))).
```

The reader must have noticed the confusion between the Lisp expressions and Lisp data as represented by the s-expression notation. For example, in the computation above, we would like to write (APPEND (1 2) (3)) for the expression being computed, but this would be incorrect. The reason is that Lisp would try to evaluate (1 2) and (3) as Lisp expressions and expect Lisp functions named 1 and 3. The QUOTE operation of Lisp is used to indicate the boundary between Lisp expression and data so that the above expression could then be correctly written as (APPEND (QUOTE (1 2)) (QUOTE (3))) (and further abbreviated as (APPEND '(1 2) '(3))). This similarity between expression and data is exploited in a crucial way by Lisp to represent Lisp expressions as data to other Lisp functions. An *interpreter* for Lisp is a program that is capable of evaluating Lisp expressions given a collection of function definitions and an *environment* associating Lisp variables to s-expression values. It is easy to write an interpreter for Lisp in Lisp itself since Lisp expressions are easily represented as s-expressions. The proofs described in this book also exploit this similarity between Lisp's notation and its s-expression data.

Lisp can be shown to have the same computing power as any of the other universal computing formalisms such as Turing machines[12] and the lambda calculus. The Boyer–Moore theorem prover (see Section 1.3 for more details) is capable of expressing and proving properties of Lisp programs such as the associativity of the APPEND function above.

1.1.9 Metamathematics

In their *Principia Mathematica*, Russell and Whitehead had shown [WR25] that a large portion of mathematics could be formally proved in a formal theory (which we call PM). Their work used no metamathematics. They proved theorems within their formal theory but not about it.

Hilbert founded metamathematics in order to demonstrate that formal theories such as PM were *consistent*, in other words, free of contradiction. If such a proof could be carried out by "safe" methods, then the foundations of mathematics were indeed secure. Hilbert had also hoped that PM could be shown to be powerful enough to prove all the true mathematical statements. His approach to these foundational problems was to treat the syntax of formal theories as the objects of his mathematical investigation.

---

[12]This fact has been mechanically verified using the Boyer–Moore theorem prover [BM84a].

The first significant results in metamathematics were Post's proofs of the decidability and completeness of propositional logic [Pos21]. The next major advance in metamathematics came with the results of Kurt Gödel. In 1931, Gödel showed that in a wide class of formal theories including PM, there existed statements that could be determined to be true of numbers but lacked any proof within the theory. This was Gödel's first incompleteness theorem. Its proof is perhaps the crowning achievement of metamathematics [Göd67b, Kle52]. Gödel also derived an important consequence of the incompleteness theorem. He showed in his second incompleteness theorem that there could be no "safe" proof of the consistency of any theory rich enough to contain some basic arithmetic. These two incompleteness theorems essentially destroyed Hilbert's programme to secure the foundations of mathematics by means of an incontrovertible proof of its consistency.

We have described metamathematics as the mathematical study of the properties of formal theories. These formal theories consist of well-formed expressions and syntactic rules for deriving expressions from other expressions. For example, first-order logic consists of expressions which are well-formed formulas, and rules of inference for deriving theorems from the axioms. The formalism of lambda calculus consists of terms, and computation rules for reducing terms. It should be clear that there are two levels of discourse in metamathematics. There is a *meta-level* at which the theorems of metamathematics will be stated and their proofs argued. The other is the *object-level* which is the formal language and theory about which metamathematical theorems (or *metatheorems*) are proved. The meta-level discourse has traditionally been carried out in informal terms in languages like English or German. In this book, however, the meta-level discourse will itself be formal. This facilitates the use of a computer for checking metamathematical proofs.

The formal *metatheory* we use will be the Boyer–Moore logic, a formalization of the programming language pure Lisp. We often refer to the Boyer–Moore logic as Lisp. The proofs were checked using the Boyer–Moore theorem prover, a theorem proving program for the Boyer–Moore logic. Two different *object theories* are considered here. One is Cohen's first-order theory of hereditarily finite sets Z2 [Coh66] shown in Section 1.1.6. The other object theory is the lambda calculus [Chu41].

1.1.10 Automatic Proof Checking

At around the same time that Gödel proved the incompleteness theorem, Herbrand [Her67] showed that there was a method to determine if a given statement was a theorem of first-order logic. The method would lead to a positive answer if the given statement was a theorem, but might fail to terminate with a definite answer when given a non-theorem. This showed that first-order logic was *semi-decidable*, and raised the theoretical possibility that computers could be programmed to prove theorems.

The Resolution Method.   Robinson's *resolution* method is a simple but powerful semi-decision procedure for first-order logic [Rob65]. At the propositional level, the resolution method works by negating the formula to be proved, and converting this negation into *conjunctive normal form* as the conjunction of the disjunction of *literals* (propositional atoms and their negations). Each such disjunction is regarded as a *clause*. Two clauses $p \vee B$ and

$\neg p \vee C$ are *resolved* to yield the new clause $B \vee C$.[13] This process is carried on until an obvious contradiction is found, indicating that the original formula is indeed a theorem.

At the first-order level, resolution works by first casting the negated conjecture into *prenex normal form* where no quantifier occurs within the scope of a propositional connective. The resulting formula has the form $(Q_1 x_1 \ldots Q_n x_n. A)$ where each Q_i is either $\forall$ or $\exists$ and A does not contain any quantifiers. This formula is then *skolemized*[14] by replacing any existential quantifiers by terms of the form $f(y_1, \ldots, y_m)$, where f is a new function symbol and the y_i are the governing universally quantified variables. We now get a formula of the form $\forall y_1 \ldots \forall y_k. A'$. The universal quantifiers are then dropped, and the resulting formula A' is placed into conjunctive normal form as the conjunction of clauses. The resolution step now uses an operation called *unification* to find a substitution σ such that $\sigma(p) \equiv \sigma(p')$ in order to resolve the clauses $p \vee B$ and $\neg p' \vee C$ to yield the clause $\sigma(B \vee C)$ (the result of applying the substitution σ to $B \vee C$). Clauses are resolved until resolution yields a contradiction. There are a number of heuristics to manage the combinatorial explosion of clauses.

The resolution method has many uses in logic programming and artificial intelligence. A great deal of effort has been directed at incorporating powerful heuristics into resolution theorem provers in order to make them more efficient [CL73]. The OTTER system [McC90, Qua90] is a powerful resolution-based theorem prover that has been used to solve a significant number and variety of nontrivial (and even open) mathematical problems. The resolution method is not directly useful for the purpose of proof checking since it cannot be easily extended with nonlogical inference techniques such as induction. The proofs generated by resolution are typically not in a style that can be easily read and understood by humans.

Other Automatic Theorem Provers. Many other researchers reacted to the weaknesses of resolution by exploring other more heuristic approaches to automatic theorem proving [Ble77]. The hope was that the proofs thus generated would be more natural and hence more conducive to human interaction. Bledsoe's IMPLY system[Ble83] is a very influential heuristic theorem prover that is based on proof search in Gentzen's sequent calculus [Sza69]. Other examples of such systems include the Rewrite Rule Laboratory (RRL) which is a powerful system based on conditional rewriting [KZ89], and the TPS system which is a theorem prover for higher-order logic based on method of *matings* [AMCP84]. PTTP is a theorem prover for first-order logic that achieves theorem proving efficiency by adapting and extending the execution framework of the programming language Prolog [Sti88]. The Boyer–Moore theorem prover [BM79, BM88] used here is another instance of a program that is not based on resolution and makes effective use of heuristics to carry out proofs by induction and rewriting.

Automatic Proof Checkers. As mentioned earlier, the formalization of the notion of a mathematical proof has made it possible for proofs to be checked by means of a computer. The Boyer–Moore theorem prover only represents one of several different attempts to real-

---

[13]This is exactly the Cut rule of the propositional logic shown in Figure 1.2 restricted to literals. The other propositional rules are only used to convert the negation of the given conjecture into a normal form and to preserve the normal form following a resolution step.

[14]Thus named in honor of the Norwegian logician Thoralf Skolem.

ize this possibility. As early as 1962, McCarthy [McC62] had strongly advocated computer proof checking as an important application of computers. He developed a mathematical theory of the programming language Lisp [McC63] that could be used to carry out proofs of properties of Lisp programs. The formal framework within which the Boyer–Moore theorem prover [BM79, BM88] proves theorems is a variant of McCarthy's theory. Another early proponent of proof checking, N. G. de Bruijn, started the AUTOMATH project to check mathematical proofs [dB70, dB80]. The project culminated in a verification of all the proofs in Landau's *Foundations of Analysis* [vBJ79]. The influence of the AUTOMATH project has been wide-ranging: several modern languages and inference systems are based on ideas originating from this project.

There are several other interesting proof checking projects. The FOL system was an experiment at using a proof checking system to reflectively prove and employ properties about itself [WT74]. The LCF (Logic for Computable Functions) system is an extensible proof checker that has been used to verify the properties of programs [GMW79, Pau87], and to check the correctness of a unification algorithm [Pau84]. The HOL system [Gor88] is a widely used descendant of LCF that, as the name suggests, formalizes higher-order logic. Nuprl [Con86] is a highly developed proof checker based on LCF that formalizes proofs in an expressive constructive type theory. Nuprl is intended as a mechanized realization of Bishop's program of reconstructing mathematics in a constructive manner. The Calculus of Constructions [CH85] is another LCF style proof checker that features an expressive type system. *Isabelle* [Pau90] is a variant of LCF that provides a higher-order logic metalanguage for defining theorem provers. λ-Prolog [MNPS91], Elf [Pfe89], and LF [HHP87] are other metalinguistic computational frameworks for defining proof checkers.

Metamathematical Extensibility.   The crucial problem with constructing and checking formal proofs is that these proofs soon become intolerably large. The LCF systems coped with this problem by allowing users to define *tactics*, programs in a metalanguage (ML) that construct formal proofs. Proofs become easier to construct by invoking tactics, but the time-consuming task of checking the formal proof remains. For example, a tactic for proving propositional tautologies would have to generate and check the proof of each tautology. The task of checking the proof generated by the tactic could be avoided by proving a metatheorem that asserts that every tautology has a correct proof and employing this as a new rule of inference. Such a rule is called a *derived inference rule*. The ability to extend a proof checker with derived inference rules was termed *metamathematical extensibility* by Davis and Schwartz [DS77].

The work reported here constitutes, to our judgement, the first convincing demonstration of a metamathematically extensible proof checker. The FOL project was an attempt in this direction, but since most metamathematical proofs would be quite tedious to check using the FOL proof checker, no significant derived inference rules were proved [WT74]. Another attempt due to Boyer and Moore uses a very restricted notion of a derived inference rule that only has limited applicability [BM81]. It is important to note that both these efforts were aimed at getting proof checkers to check extensions to themselves by *reflecting* their metatheory within the object theory. In our approach to metamathematical extensibility, we do not use reflection but instead employ the Boyer–Moore theorem prover as a metatheoretic framework for checking extensions to another proof checker that has been formalized within the Boyer–Moore logic. In particular we have extended a proof checker for first-order logic with a large number of interesting derived inference rules. These derived inference rules make it feasible to construct and check complicated formal proofs.

1.1.11 Why Check Proofs in Metamathematics?

We wish to address the question: *Can complicated mathematical proofs be checked using a computer?* This possibility has long been denied by critics of formal mathematics. We choose to examine proofs in metamathematics because it presents some classic instances of unquestionably complicated proofs. The proof of Gödel's incompleteness theorem is an example. Most informal descriptions of this proof skip many important details. This proof therefore posed a significant challenge for computer proof checking. The work described here does not demonstrate that any piece of mathematics can be mechanically proof checked within a reasonable amount of time and effort, but it does indicate that there are no obvious insurmountable obstacles to the task of verifying any individual piece of mathematics.

Apart from their depth and difficulty, proofs in metamathematics are interesting and useful because they make it feasible to extend proof checkers with powerful and correct inference rules. In the course of the proof of Gödel's incompleteness theorem, many significant derived inference rules were proved as metatheorems. Derived inference rules make it possible to check mathematical proofs at a level that is much closer to a journal-level description of the proof. These rules are also computationally efficient compared with the time-consuming process of checking the low-level formal proof.[15]

It is important to note that the above-mentioned derived inference rules are about object-level proofs and not about the metatheoretic proofs generated by the Boyer–Moore theorem prover. Though the theorem prover has a facility for similar metatheoretic extensions [BM81], it was not found useful in these proofs.

A more general question is whether it is useful to mechanically verify any mathematical proofs. Critics of computer proof checking have argued that the use of computers to check proofs subverts the social process by which mathematicians come to accept proofs [MLP79]. Computer proof checking is a strengthening of the social process and not a substitute for it. Proofs are justifications, and justifications need to be unambiguous and correct. Computers provide an effective mechanism for ensuring correctness. It would not be practical to check proofs to the desired level of detail and rigor without the use of computers. Many subtle mistakes and gaps are uncovered during formal mechanical verification that might not be as easily found by the unsupported social process.

An additional benefit of computer proof checking is that the logical structure of an argument becomes much clearer as a result and the proof can be more easily understood and communicated. The mechanical proof of the Church–Rosser theorem is a good illustration. The steps in the mechanical proof are at least as convincing and readable as previous informal descriptions of this proof. Leibniz had in fact predicted that with the development of symbolic logic, *"humanity would have a new kind of an instrument increasing the powers of reason far more than any optical instrument has ever aided the power of vision."*[16]

---

[15]Pitrat [Pit70] describes an early attempt at automatically generating some simple metatheorems of propositional logic.

[16]Quoted by Gödel in [Göd44].

1.2 Overview

This book describes the mechanical proofs of several important theorems in metamathematics. In the first part of the book we present mechanical proofs of metatheorems about an object theory consisting of the rules and axioms of Shoenfield's first-order logic [Sho67] with Cohen's Z2 axioms [Coh66]. This object theory will be referred to as Z2. The metatheorems proved include the tautology theorem, the representability theorem, and Gödel's incompleteness theorem. The second part of the book presents the highlights of a formalization and proof of the Church–Rosser theorem of the lambda calculus. The theorems were formalized in the Boyer–Moore logic and the proofs were checked using the Boyer–Moore theorem prover. These brief introductory discussions are not entirely precise and merely provide the outline for the more elaborate descriptions in the chapters to follow.

1.2.1 The Rules of the Game

Since the phrase, "to check a proof of the incompleteness theorem," is open to a wide variety of interpretations, the desired objective has to be carefully specified. The proof had to satisfy three basic requirements.

The proof had to be checked using the Boyer–Moore theorem prover with no major modifications. That is, the same prover that had previously checked proofs in number theory and program verification had to be employed. It could not be customized for the proof of the incompleteness theorem, or modified to a point where its soundness was in jeopardy. During the course of the proof attempt, some very minor modifications were made that led to improved overall performance in the theorem prover, but the logic used by the theorem prover remained unchanged.

The proof had to demonstrate an essential incompleteness in a theory which was well known and relevant to mathematical practice. More specifically, our proof had to demonstrate the incompleteness of a formal theory such as Peano arithmetic, and not a simpler theory like primitive recursive arithmetic (PRA) or quantifier-free Lisp. It is easy to make the task of proving the incompleteness theorem considerably less meaningful through the choice of the object theory. In particular, picking an object theory that is close to the metatheory trivializes the task of proving that the metatheory can be represented in the object theory. The object theory Z2 that is used here is in fact as expressive as Peano arithmetic [Coh66]. Going to the other extreme and picking a very powerful object theory like Zermelo-Fraenkel set theory would have also made the proof less challenging since this would have made it quite trivial to represent recursive functions.

Only the "safe" commands to the Boyer–Moore theorem prover could be used. This ruled out the use of function definitions whose admissibility has not been proved, the use of uninterpreted function symbols, and extra axioms. These restrictions are needed to prevent the introduction of any inconsistencies into the theorem prover's logic through the axioms.

Once the above restrictions have been followed, then in order to believe the proof, one need only

1. Believe in the soundness of the Boyer–Moore theorem prover, and

2. Be convinced that the incompleteness theorem has been correctly stated.

The remaining steps in the proof and the proof generated by the Boyer–Moore theorem prover need not be examined. Only a small number of definitions need to be examined in order to arrive at the conviction that the statement of the theorem has been correctly formulated. Chapter 2 presents the formal statement of the incompleteness theorem.

1.2.2 Steps to the Incompleteness Theorem

The development leading to the proof of Gödel's incompleteness theorem consists of a number of steps that are quite confusing. The goal is to demonstrate that the set theory Z2 is either inconsistent or incomplete. If Z2 is inconsistent in that it permits the proof of some statement and its negation, then it is easy to show that it proves every statement. Hence, the goal reduces to that of constructing a statement of the set theory Z2 that is neither provable nor disprovable in Z2, unless Z2 is inconsistent. For this purpose, the notion of a Z2 proof must be rigorously presented in the metatheory. Various familiar statements can be shown to be provable in Z2 by constructing their proofs. In particular, it is shown that Z2 is expressive enough to capture its own metatheory; there is a formula $A(p, s)$ of Z2 such that whenever s is a set-theoretic encoding of a statement of Z2 and p is a similar encoding of a valid proof of this statement, then $A(p, s)$ is provable in Z2. Otherwise, if p does not encode a valid proof of the statement encoded by s, then $\neg A(p, s)$ is provable in Z2.

Armed with this knowledge about the ability of Z2 to capture its own metatheory, we can construct a statement U in Z2 that asserts the provability of its negation $\neg U$, in Z2. If U has a Z2 proof, then it is possible to show that $\neg U$ is also provable, thus leading to an inconsistency in Z2. On the other hand, if $\neg U$ has a Z2 proof, then since $\neg U$ asserts the absence of a Z2 proof of $\neg U$, we have a provable falsehood in Z2. The only choice is either to concede that Z2 could be inconsistent or to allow for the possibility that neither U nor $\neg U$ has a proof in Z2 and that Z2 is incomplete. These steps to the incompleteness theorem are briefly described below.

1.2.3 Formalizing Z2

The object theory Z2 shown in Figures 1.2, 1.4, and 1.5 is formalized in the Boyer–Moore logic (which for the rest of this book is abbreviated as Lisp) by defining a Lisp program that checks Z2 proofs.[17] To define the proof checker, we simply view Z2 proofs as well as the symbols and expressions of Z2 as objects that can be represented using the s-expression data structure of Lisp. It is easy to see that data structures such as numbers and lists can be used to represent the syntax of Z2.

We can now define a Lisp function to check whether a given data structure represents an axiom of Z2. We can similarly define a Lisp function to check whether a given s-expression represents a well-formed proof built from the axioms and inference rules of Z2. Proofs in Z2 are represented as tree-like structures whose leaf nodes are the axioms, and whose parent nodes follow from their offspring nodes by an application of one of the rules of inference. Let Proves be the Lisp function that checks if a given data structure pf is a well-formed Z2 proof of a Z2 formula represented by exp. The Lisp expression (PROVES PF EXP) can be defined to return the value T (representing 'true') when PF does represent a valid proof of EXP, and

---

[17]Note that Lisp as formalized by the Boyer–Moore theorem prover (and as used in this book) is significantly different from popular versions of the Lisp programming language.

to return the value F (representing "false") otherwise. Chapter 2 defines a representation for Z2 proofs along with a Lisp function that checks Z2 proofs.

1.2.4 The Tautology Theorem

Having defined `Proves`, we can use it to establish the provability of certain formulas and classes of formulas in Z2. Consider the formula $A \vee B \vee \neg A$. If we assign truth values **true** or **false** to the atoms A and B, then under any assignment of truth values to A and B, the formula $A \vee B \vee \neg A$ has the value **true**. So a formula can be said to be a *tautology* if it evaluates to **true** under any assignment of the truth values, **true** or **false**, to its atoms. The *tautology theorem* asserts of Z2 that all tautologies are theorems of Z2.

We can define a Lisp function `TAUTOLOGYP` to recognize whether a given formula is a tautology so that `(TAUTOLOGYP A)` returns T exactly when the s-expression A represents a formula that is a tautology. Such a *tautology checker* can be shown to be correct with respect to the truth-table definition of a tautology. A sophisticated induction argument then establishes that a Z2 proof can be constructed for any formula recognized to be a tautology by the tautology checker. A function `TAUT-PROOF` is defined to explicitly construct the Z2 proof of a given tautology. We can show that whenever `(TAUTOLOPYP A)` is T, `(Proves (TAUT-PROOF A) A)` is also T.

The tautology theorem is an interesting labor-saving device since once we have recognized a formula to be a tautology, then we are spared the trouble of constructing its proof in Z2. The tautology theorem thus becomes a *derived inference rule* that is used to construct more concise proofs for Z2 theorems.

Numerous other examples of derived inference rules were proved sound using the Boyer–Moore theorem prover, including the instantiation rule, the equivalence theorem, and the equality theorem. The third chapter of Shoenfield's *Mathematical Logic* [Sho67] is mainly devoted to the proofs of these theorems. The mechanical proof of the tautology theorem is described in Chapter 3 of this book.

1.2.5 The Representability Theorem

The theory Z2 contains enough axioms about hereditarily finite sets to define the natural numbers and ordered pairs. The Lisp data structures can easily be represented using natural numbers and ordered pairs. Let $\lceil X \rceil$ denote the set in Z2 that encodes the Lisp data structure X. Recall that the notation $[a/x]A$ denotes "*a substituted for all free occurrences of x in A.*" More generally, let $[a_1/x_1, \ldots, a_n/x_n]A$ represent the *simultaneous* substitution of the a_i for the x_i in A. Let $\vdash A$ denote "A is provable in Z2." The *representability theorem* asserts that for any n-ary Lisp function g, there is a formula $A_{\mathbf{g}}$ with exactly $n+1$ free variables, say $y, x_1, \ldots, x_n$, such that

$$\text{If (g } X_1 \ \ldots \ X_n) = Y,$$
$$\text{then} \quad \vdash [\lceil X_1 \rceil/x_1, \ldots, \lceil X_n \rceil/x_n, \lceil Y \rceil/y]A_{\mathbf{g}}. \tag{1.1}$$

$$\text{If (g } X_1 \ \ldots \ X_n) \neq Y,$$
$$\text{then} \quad \vdash [\lceil X_1 \rceil/x_1, \ldots, \lceil X_n \rceil/x_n, \lceil Y \rceil/y](\neg A_{\mathbf{g}}). \tag{1.2}$$

The representability theorem provides the important bridge between the theory Z2 and the metatheory Lisp by showing that computations in the metatheory can be formally represented and verified in Z2. The problem of showing that *any* pure Lisp function is representable in Z2 can be reduced to that of showing that a particular Lisp function, a pure Lisp *evaluator* or *interpreter*, is representable in this manner. The Lisp evaluator computes the value of a pure Lisp term with respect to a list of bindings of values to variables, and a list of bindings of definitions to function symbols. It follows from the representability of the interpreter that any function computable using the interpreter is also representable in Z2. Derived inference rules such as the tautology theorem play a very important role in the proof of the representability theorem.

The representability theorem provides the key step in the construction of the undecidable sentence needed to prove the incompleteness theorem. Chapter 4 contains an outline of the mechanical proof of the representability theorem.

1.2.6   Gödel's Incompleteness Theorem

The statement of the incompleteness theorem for Z2 asserts the existence of an undecidable sentence U of Z2 such that if U is either provable or disprovable in Z2, then it is both provable and disprovable. Note that a formula A is disprovable if its negation $\neg A$ is provable. Since each sentence is either true or false of the intended model, a theory is *complete* only if every sentence is either provable or disprovable. It is *consistent* if no formula is both provable and disprovable. Thus the incompleteness theorem for Z2 states that Z2 is either incomplete or inconsistent.[18] Let A denote the Lisp representation of a Z2 formula A, and let (NEG A) compute the Lisp representation of $\neg A$. To prove the incompleteness theorem, we first define a Lisp function that serves as a *theorem checker* for Z2. This function, THEOREM, is defined so that:

$$\text{If (THEOREM A)} = \text{T,} \quad \text{then} \quad \vdash A. \tag{1.3}$$
$$\text{If (THEOREM A)} = \text{F,} \quad \text{then} \quad \vdash \neg A. \tag{1.4}$$

Let PR be a Lisp function that enumerates (with possible repetitions) all the Z2 proofs as (PR 0), (PR 1), (PR 2), etc. This can be done by means of an encoding of the s-expressions in terms of numbers, where PR is the inverse of this encoding. The Lisp function (THEOREM A) works by enumerating all the s-expression data structures that could possibly represent proofs as (PR 0), (PR 1), (PR 2), etc. Let (ADD1 N) be the primitive Lisp function that adds one to the number N. Then we can define THEOREM as:

```
(THEOREM X) =  (THEOREM-SEARCH X 0)

(THEOREM-SEARCH X N) =
    (IF (PROVES (PR N) X)
        T
      (IF (PROVES (PR N) (NEG X))
          F
```

---

[18]This statement of the incompleteness theorem is a strengthening of Gödel's original formulation which used the notion of ω-consistency instead of consistency. ω-consistency is a strong form of consistency where a theory is said to be ω-inconsistent if it permits a proof of both $\neg(\forall x. A(x))$ and each of $A(0), A(1)$, etc. The first proof for this strengthened form of the incompleteness theorem is due to Rosser [Ros36].

```
(THEOREM-SEARCH X (ADD1 N))))
```

In other words, (THEOREM A) searches for the first proof or disproof (i.e., proof of the negation) of a Z2 formula A (represented by the Lisp s-expression A) in the given enumeration of
possible Z2 proofs. The evaluation of (THEOREM A) need not always terminate, but it returns
either T or F when it does terminate. If, however, Z2 is assumed to be complete, then every
Z2 sentence has a proof or a disproof that will eventually be found in the enumeration.
When the computation of (THEOREM A) does terminate, the resulting value is computable
by executing the Lisp evaluator on (THEOREM A) so that we can employ the representability
theorem.

If we pick two distinguished variables x and y in Z2, and for any Z2 formula A we represent
$[s/x, t/y]A$ as $A(s,t)$, and $[s/x]A$ as $A(s)$, then we can restate (1.1) and (1.2) so that for
any Lisp function g such that the evaluation of (g X) terminates:

$$\text{If } (\text{g X}) = \text{Y}, \quad \text{then} \quad \vdash A_{\mathbf{g}}(\lceil \text{X} \rceil, \lceil \text{Y} \rceil) \tag{1.5}$$

$$\text{If } (\text{g X}) \neq \text{Y}, \quad \text{then} \quad \vdash \neg A_{\mathbf{g}}(\lceil \text{X} \rceil, \lceil \text{Y} \rceil). \tag{1.6}$$

We now need to construct the *undecidable sentence* and demonstrate that Z2 is not both
consistent and complete. Given that A is the Lisp representation for the formula A, let (SB
A) be the Lisp representation for the Z2 formula $A(\lceil A \rceil)$. The operation SB can easily be
defined as a Lisp function: it is the syntactic operation of substituting the Lisp representation
of the Z2 term $\lceil A \rceil$ for all the free occurrences of x in the Lisp representation A of the formula
A. Now if we let (g X) in (1.5) and (1.6) be (THEOREM (SB X)), and Y be T, then we have
an A (where $A(x)$ is $A_{\mathbf{g}}(x, \lceil T \rceil)$) such that

$$\text{If } (\text{THEOREM (SB X)}) = \text{T} \quad \text{then} \quad \vdash A(\lceil \text{X} \rceil) \tag{1.7}$$

$$\text{If } (\text{THEOREM (SB X)}) \neq \text{T} \quad \text{then} \quad \vdash \neg A(\lceil \text{X} \rceil). \tag{1.8}$$

If we now let G be $\neg A$ and let G be the Lisp representation of G, then the undecidable
sentence U is $G(\lceil G \rceil)$. Note that U is a sentence (i.e., it contains no free variables). Also note
that the Lisp representation of U is (SB G). We can now prove the incompleteness theorem
by showing that if Z2 is assumed to be complete, then it is inconsistent.

If Z2 is complete, the evaluation of (THEOREM (SB G)) terminates yielding either T or F.
If (THEOREM (SB G)) = T, then by (1.3), the sentence $G(\lceil G \rceil)$ is provable. But $A(\lceil G \rceil)$ is
also provable by (1.7). Since $G(\lceil G \rceil)$ is the negation of $A(\lceil G \rceil)$, this leads to an inconsistency
in Z2. If on the other hand, (THEOREM (SB G)) = F, then by (1.4), the sentence $G(\lceil G \rceil)$
is disprovable. By (1.8), the sentence $A(\lceil G \rceil)$ is also disprovable, which is to say, $G(\lceil G \rceil)$
is provable. Since $G(\lceil G \rceil)$ is both provable and disprovable, Z2 is again inconsistent. Thus
assuming Z2 to be complete leads to Z2 being inconsistent. So Z2 is either incomplete or
inconsistent, and hence the incompleteness theorem for Z2 follows.

The above argument is actually an adaptation of Berry's paradox. The sentence U can be
interpreted as asserting that (THEOREM U) returns F, that is, that the first proof of either
U or $\neg U$ in the given enumeration of proofs is that of $\neg U$. So, U is actually asserting
the provability of its own negation. Gödel's original undecidable sentence asserted its own
nonprovability and was thus based on the Liar paradox.

Note that it does not help to add U or $\neg U$ as an axiom of Z2 since the above construction can be repeated for the extended theory as well, and a new undecidable sentence can be derived. The mechanically checked proof of the incompleteness theorem consists of the definition of the Lisp function **Proves**, the proof of the representability theorem (statements (1.1) and (1.2)), and the construction of the undecidable sentence by means of a *diagonalization* argument (statements (1.7) and (1.8)). Though the diagonalization argument is the trickiest part of the informal argument, it is quite easily verified. On the other hand, the representability theorem, which is more easily believed, constitutes the major part of the mechanical verification.

1.2.7   The Church–Rosser Theorem

The second part of the book deals with the metamathematics of the lambda calculus which can be seen as a simple programming language in which the only evaluation mechanism is the replacement of formal parameters by actual ones. The Church–Rosser theorem implies the consistency of lambda calculus, that is, that two different ways of evaluating the same lambda calculus term will not yield different answers.

Church's lambda calculus [Bar78a, Chu41] provides a formalization of the notion of computability. Church [Chu36] used it to demonstrate that the decision problem for predicate calculus was unsolvable. He had actually hoped that a lambda calculus based on functions would provide an alternative to set theory as a foundation for mathematics, based on the primitive notion of operations rather than collections. This hope was soon dashed by the discovery of inconsistencies when lambda calculus was extended with logical notions. The lambda calculus has proved to be extremely fruitful in the study of the theory and implementation of programming languages.

Church observes that contrary to popular mathematical usage, the expression $x + x$ does not represent the *function* that doubles a given number; it represents the *result* of doubling x. He employed the notation $(\lambda x.\ x+x)$ to denote the function that doubles a given number. The *term* $(\lambda x.\ x+x)$ is called a *lambda abstraction*. When the lambda abstraction $(\lambda x.\ x+x)$ is applied to an argument as in $((\lambda x.\ x + x)\ 3)$, the resulting term is called a *redex*. Such a redex is evaluated by substituting the argument 3 for the free occurrences of x in the body $x + x$, of the function. The process of substituting the argument for the abstracted variable in the body of the lambda abstraction is termed *beta-reduction*. The beta-reduction of the above redex yields the expression $3 + 3$, which by the primitive operation of addition yields 6 as the result. Functions can be applied to functions (including themselves) in the lambda calculus. The function term $(\lambda f.\ (\lambda x.\ (f\ (f\ x))))$ applies the first argument f twice to its second argument x. Thus the term

$$(((\lambda f.\ (\lambda x.\ (f\ (f\ x))))\ (\lambda y.\ y + y))\ 3)$$

beta-reduces to

$$((\lambda x.\ ((\lambda y.\ y + y)\ ((\lambda y.\ y + y)\ x)))\ 3).$$

This latter term is a redex and beta-reduces to

$$((\lambda y.\ y + y)\ ((\lambda y.\ y + y)\ 3))$$

$$\overline{X \overset{\beta}{\mapsto} X} \qquad \overline{((\lambda x.\ X)\ Y) \overset{\beta}{\mapsto} [Y/x]X}$$

$$\frac{X \overset{\beta}{\mapsto} X'}{(\lambda x.\ X) \overset{\beta}{\mapsto} (\lambda x.\ X')} \qquad \frac{X \overset{\beta}{\mapsto} X' \quad Y \overset{\beta}{\mapsto} Y'}{(X\ Y) \overset{\beta}{\mapsto} (X'\ Y')}$$

$$\overline{X \longrightarrow X} \quad \frac{X \overset{\beta}{\mapsto} X'}{X \longrightarrow X'} \quad \frac{X \longrightarrow Y \quad Y \longrightarrow Z}{X \longrightarrow Z}$$

Figure 1.6: The Lambda Calculus: Beta-Steps and Reductions

which further beta-reduces to $((\lambda y.\ y + y)\ 6)$. This latter term when beta-reduced evaluates to 12. Note that there is an alternative way to carry out the beta-reductions from

$$((\lambda y.\ y + y)\ ((\lambda y.\ y + y)\ 3))$$

by beta-reducing the *outermost* redex to get

$$((\lambda y.\ y + y)\ 3) + ((\lambda y.\ y + y)\ 3)$$

then beta-reducing the two independent redexes to get $(3 + 3) + (3 + 3)$ which immediately yields 12. So these alternative sequences of beta-reductions result in the same final value.

The Church–Rosser theorem implies that this is always the case by asserting that if a term X reduces to Y through one sequence of reductions, and to Z through another sequence of reductions, then there is term W such that Y and Z can both be reduced to W. In particular, if Y and Z are values or *normal forms*, that is, they cannot be further beta-reduced, then Y and Z must be essentially identical. In simpler terms, the Church–Rosser theorem implies that different terminating evaluations of the same term yield essentially the same value. Note that certain sequences of reductions of a term might never terminate in a value. There are also terms for which no sequence of reductions will terminate in a value. The Church–Rosser theorem implies the consistency of lambda calculus [CR36].

Figure 1.6 captures the rules for a β-step so that each β-step beta-reduces zero or more non-overlapping redexes in a term. The reduction relation also shown in Figure 1.6 is just the reflexive-transitive closure of a β-step. The definitions of the substitution operation and the α-step relation which renames bound variables have been omitted.

Though proposed in 1936, the Church–Rosser theorem did not have a widely accepted proof until 1971 when it was proved by Tait and Martin-Löf. The proof involves a generalization of a β-step called a walk, where overlapping redexes can be reduced in a single walk provided the inner redex is reduced first. Though β-steps do not have the diamond property, walks do. Since a reduction can also be seen as a series of walks, we can derive the diamond property of reduction by using induction and the diamond property of walks. Thus the crux of the proof is the diamond property of walks.

The mechanical proof of the Church–Rosser theorem begins by formalizing the syntax of the lambda calculus in the Boyer–Moore logic. A form of the lambda calculus which employs the de Bruijn representation for terms is also similarly formalized. The proof of Tait and Martin-Löf has been adapted for this formalization of the lambda calculus. Finally, a translation between the de Bruijn and the standard representation is used to prove the Church–Rosser theorem for the standard formulation of the lambda calculus. These proofs are described in Chapter 6.

1.3   The Boyer–Moore Theorem Prover and its Logic

A brief description of the Boyer–Moore logic [BM79] is given below in order to make the formal details of the later chapters more comprehensible. The books *A Computational Logic* [BM79] and *A Computational Logic Handbook* [BM88] by Boyer and Moore provide a detailed description of the Boyer–Moore theorem prover and its logic.

The Boyer–Moore theorem prover [BM79, BM81, BM88] is one of the best-known heuristic theorem provers. It can be employed to prove properties of pure Lisp programs in a quantifier-free logic that is quite similar to Primitive Recursive Arithmetic [Goo64]. The theorem prover features a powerful heuristic for automatically discovering the induction scheme required for a proof. Previously proved lemmas are used to simplify terms during an attempted proof. The Boyer–Moore theorem prover has been used so far to prove a number of significant theorems in program verification [BM88], number theory [BM84c, Rus83], computability theory [BM84b], hardware verification [Hun85], and metamathematics. These theorems include: the fundamental theorem of arithmetic [BM79], Fermat's little theorem [BM84c], the pigeon-hole principle [BM84c], the unsolvability of the halting problem [BM84b], the invertibility of the RSA encryption algorithm [BM84c], Wilson's theorem [Rus83], Gauss' law of quadratic reciprocity [Rus88], the correctness of trace transformations [LH85], and the correctness of a microprocessor design [Hun85]. A special issue of the *Journal of Automated Reasoning* is devoted to articles on system verification using the Boyer–Moore theorem prover [BHMY89].

In all of the above proofs, the theorem prover does not independently generate the entire proof, but is led to the proof through a series of well thought out definitions and lemmas that are supplied by the user. Thus the Boyer–Moore theorem prover is used as a high-level proof checker.

1.3.1   The Boyer–Moore Logic

The first point to note about the logic employed by the Boyer–Moore theorem prover to construct proofs is that it employs the syntax and semantics of pure Lisp as described in Section 1.1.8. In fact the form of pure Lisp in that section is closer to the Boyer–Moore formalization of Lisp than to any widely used implementation of the Lisp programming language. The Boyer–Moore logic deals with recursive Lisp functions as long as they are *total*, that is, their evaluations terminate on all arguments. The restriction to total functions greatly simplifies the logic and its mechanization. We will frequently refer to the Boyer–Moore logic as Lisp. The second point to note about the Boyer–Moore logic is that it is based on a first-order logic with equality (as described in Section 1.1.4) but without any quantifiers. The absence of quantifiers is heavily exploited in order to raise the level of mechanization in the theorem prover. A surprisingly large fraction of discrete mathematics and computer science can still be expressed in this quantifier-free logic. The third point to note is that the logic is untyped in that the variables range over the entire domain of discourse which includes the booleans, natural numbers, lists, and other data structures.

1.3.1.1 The Basic Axioms

The axioms are stated in a quantifier-free first-order logic with equality. There are two distinct Boolean constants: (TRUE) and (FALSE) (abbreviated as T and F, respectively). The 3-place function IF is the only primitive logical operation. The term (IF X Y Z) (read informally as *If* X *then* Y *else* Z) is axiomatized below to return Z if X is equal to F, and Y otherwise.

$$X = F \;\supset\; \text{(IF X Y Z)} = Z$$
$$X \neq F \;\supset\; \text{(IF X Y Z)} = Y$$

The Lisp predicate EQUAL can be axiomatized to return T when both arguments are identical, and F otherwise.

$$X = Y \;\supset\; \text{(EQUAL X Y)} = T$$
$$X \neq Y \;\supset\; \text{(EQUAL X Y)} = F$$

The other Lisp logical operators such as OR, NOT, AND, and IMPLIES, can be defined in terms of IF as:

```
(NOT X) = (IF X F T)
(OR X Y) = (IF X T (IF Y T F))
(AND X Y) = (IF X (IF Y T F) F)
(IMPLIES X Y) = (IF X (IF Y T F) T)
```

The Boyer–Moore logic includes the axioms for the propositional connectives and equality given in Figures 1.2 and 1.4. The substitution rule of Figure 1.4 is replaced by an instantiation rule so that any instance of a theorem is also a theorem. Note that equality is the only predicate symbol in the logic. We refer to Lisp functions that only return either T or F as Lisp predicates. We next describe the shell principle for adding non-logical axioms, the principle of definition for introducing new (possibly recursive) total Lisp functions, and the induction principle.

1.3.1.2 The Shell Principle

The *shell principle* allows the user to add axioms describing inductively constructed objects. Natural numbers are axiomatized by adding a shell with a 1-place recognizer function NUMBERP, a 1-place constructor function ADD1, a 1-place destructor function SUB1, and a bottom object (ZERO). The result is an axiomatization similar to Peano's axioms for natural numbers. Abbreviations are used to represent numbers, e.g., (ZERO) is abbreviated by 0, and (ADD1 (ADD1 (ADD1 0))) is abbreviated by 3. The term (ZEROP X) returns F if X is a positive natural number, and T otherwise. For example, (ZEROP (CONS 1 2)) is T. Note that (SUB1 0) is 0.

Lists are similarly axiomatized by adding a shell with a 1-place recognizer LISTP, a 2-place constructor CONS, and two 1-place destructor functions CAR and CDR. The empty list is usually represented by the special constant NIL. Lists constructed using CONS can be abbreviated so that the list

```
(CONS (CONS 1 (CONS 2 NIL))(CONS 3 (CONS 4 (CONS 5 NIL))))
```

is abbreviated as '((1 2) 3 4 5). The expression (CONS X Y) returns the list whose first element is X and whose remaining elements form the list Y. For example, (CONS '(1 2) '(3 4 5)) returns '((1 2) 3 4 5). Conversely, (CAR X) returns the first element of a list: (CAR '((1 2) 3 4 5)) is '(1 2), and (CDR X) returns the remainder of the list: (CDR '((1 2) 3 4 5)) is '(3 4 5). Sequences of CARs and CDRs are abbreviated so that (CADDR X) abbreviates (CAR (CDR (CDR X))). The expression (LISTP (CONS 1 NIL)) and (LISTP (CONS 1 2)) both equal T whereas (LISTP NIL) and (LISTP 1) are both equal to F. Note that (CAR X) and (CDR X) are both equal to 0 when (LISTP X) is F.[19] The form (LIST s_1 ... s_n) is an abbreviation for (CONS s_1 (CONS s_2 ... (CONS s_n NIL))).

The basic theory included in the Boyer–Moore theorem prover contains shells for literal atoms (character strings), natural numbers, and lists. The domain of discourse of the Boyer–Moore logic is open-ended since new shells can always be introduced.

1.3.1.3 Recursion and Induction

The definitions of new functions are admitted as axioms only if they satisfy the *principle of definition* which ensures that the given function is *total*. Under this principle, a definition is accepted if and only if it is recursively or directly defined in terms of previously defined functions and there is some well-founded ordering on some measure of the arguments which decreases with every recursive call. A well-founded ordering is a partial ordering in which there are no infinite decreasing chains. A measure function is a Lisp function that maps the arguments of the function being defined to the domain of the well-founded ordering. Most measures return natural numbers, and the 2-place function LESSP is the standard well-founded ordering predicate on natural numbers. Every evaluation of functions admitted under this principle is guaranteed to terminate. In other words, only total functions are admitted. This ensures that no inconsistency is introduced by adding the definition as a new axiom.

The *induction principle* permits the application of any induction scheme that is justified by a well-founded ordering on some measure of the parameters (the free variables) of the theorem being proved. The termination arguments for the recursively defined functions occurring in the statement are used to justify the well-foundedness of any induction scheme that is applied in a proof.

1.3.2 The Theorem Prover

The Boyer–Moore theorem prover is a mechanization of the above logic. The commands to the theorem prover include those for adding shells, defining new functions, and proving theorems. A new function *name* with *n* arguments *arg1*, ..., *argn* and definition *body* is defined using the DEFN command as shown below:

(DEFN   *name* (*arg1* ... *argn*)   *body*)

---

[19] In most versions of Lisp, the expression (CAR 5) would cause an error on evaluation, whereas the Boyer–Moore logic requires all expressions to be *total*, that is, they must have a well-defined value. The choice of the value 0 rather than NIL might also seem somewhat strange. The reasons for this choice are internal to the theorem prover and not easily explained.

For example, the definition of APPEND from page 15 would be supplied to the theorem prover as:

```
(DEFN APPEND (X Y)
   (IF (LISTP X)
       (CONS (CAR X)(APPEND (CDR X) Y))
       Y))
```

The theorem prover tries to establish that the given definition is admissible by finding a measure and a well-founded ordering such that the measure of the arguments in each recursive call is smaller than the measure of the arguments to the function. In the case of APPEND, the size of the first argument decreases according to the LESSP ordering. The theorem prover also notes some simple facts that are obvious from the definition of APPEND, namely that APPEND either returns a list or returns its second argument Y.

To start the theorem prover off on the proof of a lemma *lemma-name* of type *lemma-type* which is stated as *statement*, the PROVE-LEMMA command is used as follows:

> (PROVE-LEMMA   *lemma-name*   (*lemma-type*)   *statement*)

For example, in order to prove the associativity of APPEND, the theorem prover is given:

```
(PROVE-LEMMA ASSOC-OF-APPEND (REWRITE)
   (EQUAL (APPEND (APPEND X Y) Z)
          (APPEND X (APPEND Y Z))))
```

PROVE-LEMMA takes an optional fourth argument by means of which an induction or the use of a lemma can be suggested to the theorem prover. The lemma-type indicates how the lemma is to be used in future proofs. The most common lemma type is REWRITE, which indicates that the lemma is to be used as a conditional rewrite rule. The theorem prover is automatic in the sense that once PROVE-LEMMA has been invoked on a conjecture, the user may no longer interfere with the proof. The user can, however, "train" the theorem prover by means of a carefully selected sequence of definitions and lemmas, thus directing it to check a particular proof. This is the sense in which the theorem prover is used as a high-level proof checker.

When given a conjecture, the theorem prover first tries to prove the theorem by means of *simplification* and rewriting using the axioms and previously proved lemmas. If the proof does not succeed at this stage then we are left with one or more subgoals. The steps described below are used to process each remaining subgoal. After each such step, any new subgoals are once again subject to simplification. Following simplification, the theorem prover tries to trade destructor terms like (CAR X) and (CDR X) for constructor terms by replacing X with (CONS U V), (CAR X) with U, and (CDR X) with V. This *elimination* step introduces variables so that it is possible to formulate induction schemes over these variables. The next *cross-fertilization* step applies any antecedent equalities to replace one term by another in the rest of the formula and then discards these equalities. The *generalization* step tries to replace terms in a goal formula with variables, with the expectation that an induction over the newly introduced variables might be necessary. The above steps could leave some irrelevant subterms in the goal formula that are discarded. Induction is applied to any subgoals that are not proved by the above steps. There is an induction heuristic that analyzes the recursion schemes of recursive functions in a goal formula in order to synthesize a suitable induction scheme. The first application of induction is to the original conjecture rather than to a simplified subgoal. To use the theorem prover effectively, one must be somewhat knowledgeable about the internal mechanisms and heuristics employed by the system.

1.3.2.1   A Sample Proof

We now present a sample proof carried out using the Boyer–Moore theorem prover to illustrate the above overview. Initially, the theorem prover already contains the axiomatization of the booleans, the natural numbers, and lists. It also contains the definitions of addition and multiplication operations PLUS and TIMES, respectively, and various facts about such operations. The definition of APPEND has already been shown on page 15. We first define the operation SUM that sums all the elements in a list. The given list can contain non-numbers which are simply treated as 0. When given the recursive definition of SUM, the prover first checks that the recursion is guaranteed to terminate and then accepts the definition as an axiom while printing out the commentary shown below.

```
(DEFN SUM (L)
 (IF (LISTP L)
     (PLUS (CAR L)
           (SUM (CDR L)))
     0))
```

Linear arithmetic and the lemma CDR-LESSP inform us that the measure (COUNT L) decreases according to the well-founded relation LESSP in each recursive call. Hence, SUM is accepted under the definitional principle. From the definition we can conclude that (NUMBERP (SUM L)) is a theorem.


```
[ 0.1 0.3 0.2 ]
SUM
```

Next, we try to demonstrate that the sum of the result of appending two lists X and Y can be obtained by adding the sums over X and Y, respectively. The commentary automatically generated by the prover is shown below and it is largely self-explanatory.

```
(PROVE-LEMMA SUM-APPEND
             (REWRITE)
             (EQUAL (SUM (APPEND X Y))
                    (PLUS (SUM X) (SUM Y))))
```

Call the conjecture *1.

Perhaps we can prove it by induction. Three inductions are suggested by terms in the conjecture. They merge into two likely candidate inductions. However, only one is unflawed. We will induct according to the following scheme:

```
(AND (IMPLIES (AND (LISTP X) (p (CDR X) Y))
```

```
                    (p X Y))
          (IMPLIES (NOT (LISTP X)) (p X Y))).
```
Linear arithmetic and the lemma CDR-LESSP can be used to prove that the
measure (COUNT X) decreases according to the well-founded relation LESSP in
each induction step of the scheme. The above induction scheme leads to two
new goals:

```
Case 2. (IMPLIES (AND (LISTP X)
                      (EQUAL (SUM (APPEND (CDR X) Y))
                             (PLUS (SUM (CDR X)) (SUM Y))))
                 (EQUAL (SUM (APPEND X Y))
                        (PLUS (SUM X) (SUM Y)))),
```

which simplifies, applying the lemmas COMMUTATIVITY-OF-PLUS, CDR-CONS,
CAR-CONS, and ASSOCIATIVITY-OF-PLUS, and opening up the definitions of
APPEND and SUM, to:

```
     T.
```

```
Case 1. (IMPLIES (NOT (LISTP X))
                 (EQUAL (SUM (APPEND X Y))
                        (PLUS (SUM X) (SUM Y)))),
```

which simplifies, unfolding the functions APPEND, SUM, EQUAL, and PLUS, to:

```
     T.
```

```
     That finishes the proof of *1.  Q.E.D.
```

```
[ 0.0 3.6 0.8 ]
SUM-APPEND
```

Proceeding similarly, one can introduce complex definitions and prove complicated theorems by making sure that these are broken into pieces that can be handled automatically by the theorem prover.

This concludes the overview of the Boyer–Moore logic. Other details will be provided when needed to interpret the input to the theorem prover.

1.4   Summary

We have argued in this chapter that:

- Proofs constitute the central foundational concept of mathematics and computing.

- The paradoxes suggest that the notion of a mathematical proof needs to be rigorously defined by means of a formal system. Several formal systems including propositional logic, first-order logic, Peano arithmetic, set theory (Z2), and lambda calculus were presented.

- Computers can be used to assist in the development of rigorous proofs. There are several approaches to the mechanization of proof construction ranging from the resolution and non-resolution approaches to the automated checking of formal, machine-generated proofs.

- The Boyer–Moore theorem prover proves theorems about recursive Lisp functions in a quantifier-free first-order logic. It employs induction, simplification, rewriting, and generalization in attempting to prove a given theorem.

- The metatheory of formal theories such as Z2 can be formalized using the Boyer–Moore theorem prover.

- Proofs of significant metatheorems such as the tautology theorem, the representability theorem, Gödel's incompleteness theorem, and the Church–Rosser theorem can be constructed with the aid of the Boyer–Moore theorem prover.

The remaining chapters provide the details of the mechanical verification of the above-mentioned proofs.

Chapter 2

The Statement of the Incompleteness Theorem

> *It is well known that the development of mathematics in the*
> *direction of greater precision has led to the formalization*
> *of extensive mathematical domains, in the sense that proofs*
> *can be carried out according to a few mechanical rules.*
>
> Kurt Gödel [Göd67b]

Thus began Gödel's 1931 paper [Göd67b] in which he demonstrated the existence of formally undecidable sentences in a wide class of formal systems. The first part of this book describes a formalization and proof of this theorem that was carried out according to a "few mechanical rules." The proof here is not a direct mechanization of any particular previously published version of the incompleteness proof. The statement of the theorem differs from Gödel's original statement which involved the notion of ω-*consistency*. The statement here only involves the weaker notion of *consistency* and this form of the theorem was first proved by Rosser [Ros36]. The theorem we establish asserts the incompleteness of Cohen's Z2 [Coh66]. The first-order logic of Z2 is taken from that of Shoenfield [Sho67]. Various metatheorems about Z2 are formalized and proved using the Boyer–Moore theorem prover.

This chapter presents a complete description of the formal statement of the incompleteness theorem. The main part of this description is the definition of the metatheory of Z2 in terms of a Lisp representation for the formulas and proofs of Z2, and a Lisp predicate that checks the validity of Z2 proofs represented in this manner. The representation of the symbols (variables, functions, predicates) and expressions (terms, atomic formulas, negation, disjunction, quantification) is first described. Various operations are defined for checking the well-formedness of expressions along with the important operation of substitution. A representation for Z2 proofs is given and a proof checker for these proofs is defined. This Lisp predicate for checking proofs is used to formulate the statement of the incompleteness theorem for Z2. This theorem asserts the existence of an undecidable sentence U of Z2 such that if U is provable or disprovable in Z2, then it is both provable and disprovable. Thus the completeness of Z2 implies its inconsistency.

In order to be completely rigorous, we will be presenting the Lisp definitions as given to the theorem prover but prefacing them with informal explanations in more conventional notation. Section 2.1 describes the construction in Lisp of a proof checker for Z2. Section 2.2 uses the definition of the proof checker for Z2 to formally state the incompleteness theorem.

2.1   The Metatheoretic Description of Z2

This section is devoted to the formal, metatheoretic description of the object theory Z2 in Lisp. Annotations are provided to make the pure Lisp definitions comprehensible. Section 2.1.1 introduces the *symbols* of the formal logic that will be used to construct *expressions* in Section 2.1.2. Some of these expressions are well-formed and we will define Lisp predicates[1] that recognize well-formed expressions. Section 2.1.3 describes several operations on expressions, the most important of which is *substitution*. Prior to defining the actual proof checker (in Section 2.1.8), we present Lisp predicates that recognize the Lisp representations of the logical axioms of Z2 (in Section 2.1.4), applications of the logical rules of inference (in Section 2.1.5), non-logical axioms of Z2 (in Section 2.1.6), and admissible function and predicate definitions (in Section 2.1.7). Section 2.1.9 contains an example of a Lisp data structure representing a Z2 proof that has been checked by the Z2 proof checker. The object theory Z2 used here contains a few basic function and predicate definitions given in Section 2.1.10 that have been checked to conform to the admissibility constraints given in Section 2.1.7.

To help understand the formal description of the metatheory of Z2 in Lisp, a parallel informal description of the metatheory of Z2 is also given. The metalinguistic conventions for this informal presentation are worth noting. Unsubscripted versions of x and subscripted and unsubscripted versions of y and z are used as metavariables ranging over the symbols representing the variables of the object theory. The subscripted symbols x_i for some i are used to informally represent the variables themselves. Subscripted and unsubscripted versions of the metavariables a, b, and c range over the terms of Z2. The subscripted symbols of the form f_i are used to informally represent the function symbols of Z2, and similarly subscripted symbols of the form p_i are used to represent the predicate symbols of Z2. The metavariables f and g range over function symbols, and p and q range over the predicate symbols. The metavariables A, B, and C range over the formulas of Z2. The use of distinct metavariables does not mean that the expressions instantiating these metavariables are distinct. For example, when we write $(\forall x. (\exists y. A \vee B))$, the object variables instantiating x and y need not be distinct, and similarly for the formulas represented by A and B. When such distinctness is required, as in Chapters 4 and 5, it will be explicitly indicated. In reading the Lisp definitions, it is important to keep in mind that the Lisp logic used here has no types: the variables range over the entire universe of Lisp data in the metatheory. For this reason, it will be necessary to define and use a number of Lisp recognizer predicates for the different classes of data structures in this universe.

2.1.1   The Symbols

As presented in Shoenfield's book [Sho67], the language of first-order logic consists of:

1. Denumerably many individual variables: $x_0, x_1, \ldots$;

2. Function symbols of the form f_i^j, where i is the index and j is the arity;

---

[1] Recall that Lisp predicates are Lisp functions that return either T or F.

Z2 expression	Lisp representation
x_i	i
f_i^j	(CONS 2 (CONS i j))
p_i^j	(CONS 3 (CONS i j))
$\neg A$	(F-NOT A)
$A \vee B$	(F-OR A B)
$(\exists x.\ A)$	(FORSOME X A)

Figure 2.1: Lisp Representations for Z2 Expressions

3. Predicate symbols of the form p_i^j, where i is the index and j is the arity, and $=$ is the dyadic predicate representing equality;

4. Logical connectives: $\neg$ (logical negation), and $\vee$ (logical disjunction); and

5. The existential quantifier: $\exists$.

The Lisp representations for the syntax of Z2 are shown in Figure 2.1. The Lisp data structures representing Z2 expressions are labelled *z-expressions*. The denumerably many variables are simply represented by the Lisp numerals 0, (ADD1 0), (ADD1 (ADD1 0)), and so on. A function symbol f_i^j with index i and arity j is represented by the CONS pair of the form (CONS 2 (CONS i j)). A predicate symbol p_i^j is similarly represented by a CONS pair of the form (CONS 3 (CONS i j)). The operation of negation, ($\neg A$), is represented by a new shell (F-NOT A). Logical disjunction, ($A \vee B$), is represented by another shell (F-OR A B). Existential quantification, ($\exists x.\ A$), is also represented by a shell (FORSOME X A).

The sequence of *events* (i.e., shell introduction, definitions, and lemmas) begins with BOOT-STRAP loads up the basic library of axioms about natural numbers, lists, and literal atoms.

```
    (BOOT-STRAP)
```

FUNCTION is the recognizer for object-theoretic function symbols, which have to be of the form (CONS 2 X).

```
    Definition.
      (FUNCTION FN)
      =
      (EQUAL (CAR FN) 2)
```

VARIABLE recognizes object-theoretic variables which are represented by the natural numbers.

```
    Definition.
      (VARIABLE X)
      =
      (NUMBERP X)
```

The recognizer PREDICATE for object-theoretic predicate symbols requires its argument to have the form (CONS 3 X), where X contains the index and the arity.

```
Definition.
  (PREDICATE P)
    =
  (EQUAL (CAR P) 3)
```

The DEGREE of a function or predicate symbol is its arity. These symbols are of the form (CONS n (CONS INDEX ARITY)), where n is 2 or 3. (CDDR FN) is (CDR (CDR FN)). The built-in function FIX is defined so that (FIX X) returns X if X is a number, and 0 otherwise, that is, it *coerces* its argument to a number.

```
Definition.
  (DEGREE FN)
    =
  (FIX (CDDR FN))
```

INDEX returns the index (subscript) of a function or predicate symbol.

```
Definition.
  (INDEX FN)
    =
  (FIX (CADR FN))
```

(V X) constructs the variable with subscript X.

```
Definition.
  (V X)
    =
  (FIX X)
```

(FN X Y) generates a function symbol f_X^Y with index X and arity Y.

```
Definition.
  (FN X Y)
    =
  (CONS 2 (CONS (FIX X) (FIX Y)))
```

(P X Y) generates a predicate symbol p_X^Y with index X and arity Y.

```
Definition.
  (P X Y)
    =
  (CONS 3 (CONS (FIX X) (FIX Y)))
```

The above definitions introduce the metatheoretic definitions for the variables, function symbols, and predicate symbols of Z2. We can now see how expressions involving the propositional connectives and quantifiers are constructed. Three new shells are introduced to represent the formal operations of negation, disjunction, and existential quantification, respectively.[2]

The shell constructor F-NOT constructs the expression (F-NOT X) representing the negation of a given expression X. The 'F-' prefix stands for 'formal' and is there to distinguish it from the logical NOT operation in the Boyer–Moore logic. This distinction is important: (F-NOT A) is an object in the Boyer–Moore logic which represents the object-theoretic expression $(\neg A)$, whereas (NOT X) is an expression of the Boyer–Moore logic that applies the negation operation to X. (F-NOTP X) is T if X is of the form (F-NOT Y), and F otherwise. (ARG X) is Y if X is of the form (F-NOT Y), and 0 otherwise.

```
Shell Definition.
  Add the shell F-NOT of one argument with
  recognizer F-NOTP,
  accessor ARG,
  with no type restrictions,
  and default value ZERO.
```

Similarly, (F-OR A B) is the object that represents the object-theoretic expression $(A \vee B)$. The destructors ARG1 and ARG2 return the first and second disjuncts, respectively. Negation ($\neg$) and disjunction ($\vee$) are the two basic propositional connectives.[3] The other connectives will be defined in terms of F-NOT and F-OR.

```
Shell Definition.
  Add the shell F-OR of two arguments with
  recognizer F-ORP,
  accessors ARG1 and ARG2,
  with no type restrictions,
  and default values ZERO and ZERO.
```

(FORSOME X A) represents the object-theoretic expression $(\exists x.\ A)$. The X in (FORSOME X A) is always coerced to a number, that is, a variable in the object theory so that (BIND (FORSOME X A)) returns X, or 0 when X is not a number. (BODY (FORSOME X A)) returns A. It is important to note that FORSOME is purely a syntactic operation which generates a z-expression.

```
Shell Definition.
  Add the shell FORSOME of two arguments with
  recognizer FORSOMEP,
  accessors BIND and BODY,
  with the first argument restricted to type  NUMBERP,
  no type restrictions on the second argument,
  and default values ZERO and ZERO.
```

We have so far seen the Lisp representation of the symbols in the syntax of Z2. We next look at how the expressions, well-formed formulas, and proofs of Z2 are represented.

---

[2]The reason these operations are not represented using lists and natural numbers is because it is important to know that an expression can be of at most one of these three types. The fact that distinct shells are disjoint is built into the theorem prover and is used very effectively during simplification. An explicit statement of the disjointness of classes of z-expressions would be much less effectively used by the theorem prover.

[3]We will often omit parentheses to improve the readability of the informal presentation of Z2 expressions when the meaning is clear. Note that negation binds the tightest of the propositional connectives, and disjunction associates to the right.

2.1.2   The Expressions

The next step is to show how the symbols defined in the previous section can be combined to form Z2 expressions. Only some of these expressions are well-formed terms or formulas. A *term* is either a variable or an *n*-ary function symbol followed by a list of *n* smaller terms. An *n*-ary predicate symbol followed by a list of *n* terms is an *atomic formula*. A *formula* in Z2 is either an atomic formula, a negation of a smaller formula, a disjunction of two smaller formulas, or the existential quantification of a variable over a smaller formula.

The next concept we define is a notion of equality on expressions and this definition anticipates an important need in the later parts of the proof. We cannot directly use the Lisp equality predicate, EQUAL, for checking equality of expressions since we intend to eventually Gödel-encode objects representing expressions through an encoding for all z-expressions. While we can define unique encodings for each of the objects representing expressions, the Boyer–Moore theorem prover permits the introduction of new objects to the universe through the shell mechanism; all such new objects will have to be coerced to some common default Gödel-encoding. In this case, there could be two objects that are not EQUAL in the Boyer–Moore logic but have the same Gödel-encoding. Therefore there can be no object-theoretic counterpart to EQUAL on the encoding. So in the metatheoretic descriptions to follow, we use the predicate EQL instead of EQUAL. The EQL predicate essentially checks if two z-expressions are the same modulo the coercion of those subterms which are not constructed by the shells delineated above, to the default NIL. In cases when we know that EQL and EQUAL behave identically, we prefer to use EQUAL since the theorem prover only applies rewrite rules relative to the EQUAL equality relation.

The function FIX1 coerces a non-NUMBERP argument to NIL, and is used in the definition of EQL below.

```
Definition.
  (FIX1 X)
     =
  (IF (NUMBERP X) X NIL)
```

The purpose of the coarser equality predicate EQL has already been outlined above. It contains a recursive case corresponding to each of the shells corresponding to the recognizers LISTP, F-NOTP, F-ORP, and FORSOMEP. In the base case, EQL uses FIX1 and EQUAL to check for equality.

```
Definition.
  (EQL X Y)
    =
  (IF
    (LISTP X)
    (IF (LISTP Y)
        (AND (EQL (CAR X) (CAR Y))
             (EQL (CDR X) (CDR Y)))
        F)
    (IF (LISTP Y)
        F
        (IF (F-NOTP X)
            (IF (F-NOTP Y)
                (EQL (ARG X) (ARG Y))
                F)
            (IF (F-NOTP Y)
                F
                (IF (F-ORP X)
                    (IF (F-ORP Y)
                        (AND (EQL (ARG1 X) (ARG1 Y))
                             (EQL (ARG2 X) (ARG2 Y)))
                        F)
                    (IF (F-ORP Y)
                        F
                        (IF (FORSOMEP X)
                            (IF (FORSOMEP Y)
                                (AND (EQUAL (BIND X) (BIND Y))
                                     (EQL (BODY X) (BODY Y)))
                                F)
                            (IF (FORSOMEP Y)
                                F
                                (EQUAL (FIX1 X)
                                       (FIX1 Y)))))))))))
```

The membership predicate MEMB as used in (MEMB X Y) checks if some Z such that (EQL X Z) is T, occurs in the list Y.

```
Definition.
  (MEMB X Y)
    =
  (IF (LISTP Y)
      (IF (EQL X (CAR Y))
          T
          (MEMB X (CDR Y)))
      F)
```

SYMB checks if X is a function or predicate symbol that is either the predicate symbol for equality, given by (P 0 2), or one that appears in a given list of symbols SYMBOLS. It is used to check if the function and predicate symbols used in an expression are part of the given first-order language.

```
Definition.
  (SYMB X SYMBOLS)
    =
  (OR (EQUAL X (P 0 2))
      (MEMB X SYMBOLS))
```

TERMP is the recognizer for expressions which represent *terms* in the object theory Z2. Since a term is either a variable or an *n*-ary function symbol (which appears in SYMBOLS) followed by a list of *n* smaller terms, TERMP has to be defined to recognize both single terms and lists of terms. Normally, this would be defined by means of two mutually recursive functions, but mutually recursive definitions are not directly admissible in the Boyer–Moore logic. Two or more mutually recursive definitions can be presented as a single definition that takes an extra argument to distinguish the functions. For example, the function TERMP takes an additional argument FLG to simulate the mutual recursion. When (ZEROP FLG) holds, it checks whether EXP is a single term. Otherwise, it checks if EXP is a list of terms. If (ZEROP FLG) is T, then TERMP checks that EXP is either a variable, or a function symbol (occurring in the list SYMBOLS) followed by a *list* of terms of length equal to the arity (or DEGREE) of the function symbol. If (ZEROP FLG) is F, TERMP checks that EXP is either empty, or is a single term followed by a smaller list of terms. In several of the definitions to follow, the additional argument FLG is often used in this way to simulate mutual recursion. (NLISTP EXP) abbreviates (NOT (LISTP EXP)).

```
Definition.
  (TERMP EXP FLG SYMBOLS)
    =
  (IF (ZEROP FLG)
      (IF (LISTP EXP)
          (AND (FUNCTION (CAR EXP))
               (SYMB (CAR EXP) SYMBOLS)
               (TERMP (CDR EXP) 1 SYMBOLS)
               (EQUAL (LENGTH (CDR EXP))
                      (DEGREE (CAR EXP))))
          (VARIABLE EXP))
      (IF (NLISTP EXP)
          T
          (AND (TERMP (CAR EXP) 0 SYMBOLS)
               (TERMP (CDR EXP) 1 SYMBOLS))))
```

The Lisp predicate ATOMP is the recognizer for an *atomic formula*, that is, a formula which consists of an *n*-ary predicate symbol from SYMBOLS followed by a list of *n* terms.

```
Definition.
  (ATOMP EXP SYMBOLS)
    =
  (AND (PREDICATE (CAR EXP))
       (SYMB (CAR EXP) SYMBOLS)
       (EQUAL (LENGTH (CDR EXP))
              (DEGREE (CAR EXP)))
       (TERMP (CDR EXP) 1 SYMBOLS))
```

The Lisp predicate FORMULA is the recognizer for expressions representing formulas in the object theory Z2. A formula is either of the form (F-NOT A), where A is a formula; of the form (F-OR A B), where A and B are formulas; of the form (FORSOME X A), where A is a formula and X is coerced to be a formal variable (by the shell-constructor FORSOME); or an atomic formula. The recognizer FORMULA is used in the statement of the incompleteness theorem in page 69.

```
Definition.
  (FORMULA EXP SYMBOLS)
    =
  (IF (F-NOTP EXP)
      (FORMULA (ARG EXP) SYMBOLS)
      (IF (F-ORP EXP)
          (AND (FORMULA (ARG1 EXP) SYMBOLS)
               (FORMULA (ARG2 EXP) SYMBOLS))
          (IF (FORSOMEP EXP)
              (FORMULA (BODY EXP) SYMBOLS)
              (ATOMP EXP SYMBOLS))))
```

Having defined the syntax of Z2 expressions, we can now define several important operations on expressions dealing with substitution, and free and bound variables.

2.1.3   Operations on Expressions

The main operation on expressions is that of substituting a term for a variable in an expression. In carrying out such a substitution, the free variable occurrences in the term must not become bound in the result of the substitution. The operations associated with substitution are defined below.

The recognizer (FUNC-PRED X) checks if X is a function or a predicate symbol.

```
Definition.
  (FUNC-PRED X)
    =
  (OR (FUNCTION X) (PREDICATE X))
```

The function APPEND concatenates two lists.

```
Definition.
  (APPEND X Y)
    =
  (IF (LISTP X)
      (CONS (CAR X) (APPEND (CDR X) Y))
      Y)
```

The function DEL deletes all occurrences of X from the list Y.

```
Definition.
  (DEL X Y)
    =
  (IF (LISTP Y)
      (IF (EQL X (CAR Y))
          (DEL X (CDR Y))
          (CONS (CAR Y) (DEL X (CDR Y))))
      Y)
```

The function COLLECT-FREE performs the important operation of collecting all the free variables of a given expression into a list. If (ZEROP FLG) holds, the argument EXP is a single expression. In this case, there are several cases to the definition according to whether EXP is a variable, a negation, a disjunction, an existential quantification, or a function or predicate symbol applied to a list of terms. If EXP is a variable, it is clearly free, so the list containing EXP as its only element is returned. In the case that EXP is a negation, COLLECT-FREE returns the result of collecting the free variables in the subformula of which EXP is a negation. If EXP is a disjunction, the result is got by concatenating the two lists of free variables returned from each of the disjuncts. In the case when EXP is of the form ($\exists x.\ A$), COLLECT-FREE returns the result of deleting all occurrences of x from the list of free variables in A, since all those occurrences of x are now bound. In the remaining case when the expression is a function or predicate symbol followed by a list of terms, the concatenation of the lists of free variables from each of the terms is returned. Since this last recursive invocation of COLLECT-FREE is on a list of expressions, FLG takes the value 1. When EXP is a list of expressions, (ZEROP FLG) is F, and COLLECT-FREE builds up the list obtained by concatenating the lists of the free variables from each of the members of EXP. For example,

```
(COLLECT-FREE (FORSOME 1 (LIST (P 0 3) 0 1 2)) 0)
```

returns the list of free variables (0 2), where (FORSOME 1 (LIST (P 0 3) 0 1 2)) is the z-expression representing ($\exists x_1.\ p_0^3(x_0, x_1, x_2)$).

```
Definition.
  (COLLECT-FREE EXP FLG)
    =
  (IF (ZEROP FLG)
      (IF (VARIABLE EXP)
          (CONS EXP NIL)
          (IF (F-NOTP EXP)
              (COLLECT-FREE (ARG EXP) 0)
              (IF (F-ORP EXP)
                  (APPEND (COLLECT-FREE (ARG1 EXP) 0)
                          (COLLECT-FREE (ARG2 EXP) 0))
                  (IF (FORSOMEP EXP)
                      (DEL (BIND EXP)
                           (COLLECT-FREE (BODY EXP) 0))
                      (IF (LISTP EXP)
                          (COLLECT-FREE (CDR EXP) 1)
                          NIL)))))
      (IF (LISTP EXP)
          (APPEND (COLLECT-FREE (CAR EXP) 0)
                  (COLLECT-FREE (CDR EXP) 1))
          NIL))
```

A formula is a sentence and satisfies the predicate SENTENCE if it contains no free variables.

```
Definition.
  (SENTENCE EXP)
    =
  (EQUAL (COLLECT-FREE EXP '0) 'NIL)
```

The function COVERING returns a list consisting of all those variables x such that EXP contains a subexpression of the form ($\exists x.\ A$) where A contains an occurrence of the variable VAR that is free in EXP. In other words, it returns the list of bound variables whose scopes contain free occurrences of VAR. The recursion is similar to that of COLLECT-FREE. Note that COVERING looks for bound variables in subterms of EXP, even though well-formed terms do not contain bound variables. For example,

```
(COVERING (FORSOME 1 (LIST (P 0 3) 0 1 2)) 2 0)
```

returns the list of bound variables (1) since 1 represents the only bound variable that is bound by a quantifier that *governs* a free occurrence of the variable 2 in the given expression.

```
Definition.
   (COVERING EXP VAR FLG)
     =
  (IF (ZEROP FLG)
      (IF (F-NOTP EXP)
          (COVERING (ARG EXP) VAR 0)
          (IF (F-ORP EXP)
              (APPEND (COVERING (ARG1 EXP) VAR 0)
                      (COVERING (ARG2 EXP) VAR 0))
              (IF (FORSOMEP EXP)
                  (IF (MEMB VAR (COLLECT-FREE EXP 0))
                      (CONS (BIND EXP)
                            (COVERING (BODY EXP) VAR 0))
                      NIL)
                  (IF (LISTP EXP)
                      (COVERING (CDR EXP) VAR 1)
                      NIL))))
      (IF (LISTP EXP)
          (APPEND (COVERING (CAR EXP) VAR 0)
                  (COVERING (CDR EXP) VAR 1))
          NIL))
```

NIL-INTERSECT checks that the lists X and Y have no members in common.

```
Definition.
   (NIL-INTERSECT X Y)
     =
  (IF (LISTP X)
      (AND (NOT (MEMB (CAR X) Y))
           (NIL-INTERSECT (CDR X) Y))
      T)
```

A term TERM is said to be *free for* the variable VAR in the expression EXP if no free occurrence of VAR in EXP is governed by a quantifier that binds some variable which occurs free in TERM. This is the definition of *substitutability*, since it implies that if TERM is substituted for all free occurrences of VAR in EXP, then no free variable in TERM becomes bound in the result. This definition of substitutability is captured by the predicate FREE-FOR below.

```
Definition.
  (FREE-FOR EXP VAR TERM FLG)
    =
  (NIL-INTERSECT (COVERING EXP VAR FLG)
                 (COLLECT-FREE TERM 0))
```

The substitution operation SUBST is defined by recursion on the structure of the expression EXP. It returns the result of replacing all the free occurrences of the variable VAR in EXP by TERM. This will be informally represented as $[a/x]A$, where A is the EXP, x is the VAR, and a is the TERM.

```
Definition.
  (SUBST EXP VAR TERM FLG)
    =
  (IF (ZEROP FLG)
      (IF (VARIABLE EXP)
          (IF (EQUAL EXP VAR) TERM EXP)
          (IF (F-NOTP EXP)
              (F-NOT (SUBST (ARG EXP) VAR TERM 0))
              (IF (F-ORP EXP)
                  (F-OR (SUBST (ARG1 EXP) VAR TERM 0)
                        (SUBST (ARG2 EXP) VAR TERM 0))
                  (IF (FORSOMEP EXP)
                      (IF (EQL (BIND EXP) VAR)
                          EXP
                          (FORSOME (BIND EXP)
                                   (SUBST (BODY EXP)
                                          VAR
                                          TERM
                                          0)))
                      (IF (LISTP EXP)
                          (CONS (CAR EXP)
                                (SUBST (CDR EXP) VAR TERM 1))
                          EXP)))))
      (IF (LISTP EXP)
          (CONS (SUBST (CAR EXP) VAR TERM 0)
                (SUBST (CDR EXP) VAR TERM 1))
          EXP))
```

Whenever the substitution operation $[a/x]A$ is used below, we need to ensure that a is free for x in A. This constraint on substitution could have been made part of its definition, but the definition of FREE-FOR would still be employed in the lemmas on substitution. For example, using SUBST to replace x_0 in $(\exists x_1.\ p_0^3(x_0, x_1, x_2))$ by x_1 yields $(\exists x_1.\ p_0^3(x_1, x_1, x_2))$. Note how the first occurrence of x_1 in the latter expression is *captured*, that is, it inadvertently becomes bound. On the other hand, if we apply SUBST to replace x_1 in $(\exists x_1.\ p_0^3(x_0, x_1, x_2))$ by x_2, we get $(\exists x_1.\ p_0^3(x_0, x_1, x_2))$ back unchanged since there are no free occurrences of x_1 in the expression.

There are a few other operations on z-expressions that we need in order to define the theory Z2. Many of these operations construct z-expressions that represent equality, implication, conjunction, universal quantification, etc. The dyadic predicate symbol of index 0 is used to represent equality in the object theory. (F-EQUAL X Y) returns the expression representing $(x = y)$.

```
Definition.
   (F-EQUAL X Y)

     =

   (LIST (P O 2) X Y)
```

The predicate `VAR-LIST` checks if `LIST` is a list of variables of length `N`.

```
Definition.
   (VAR-LIST LIST N)

     =

   (IF (OR (ZEROP N) (NLISTP LIST))
       (AND (ZEROP N) (NLISTP LIST))
       (AND (VARIABLE (CAR LIST))
            (VAR-LIST (CDR LIST) (SUB1 N))))
```

The function `SUB` (for 'subset') checks if all the elements of the list `X` are contained in the list `Y`.

```
Definition.
   (SUB X Y)

     =

   (IF (LISTP X)
       (AND (MEMB (CAR X) Y) (SUB (CDR X) Y))
       T)
```

`(F-AND A B)` constructs the expression representing $(A \wedge B)$ in the object theory. It is defined in terms of `F-NOT` and `F-OR`.

```
Definition.
   (F-AND X Y)

     =

   (F-NOT (F-OR (F-NOT X) (F-NOT Y)))
```

`(F-IMPLIES A B)` constructs the expression representing $(A \supset B)$, and is defined in terms of `F-OR` and `F-NOT`.

```
Definition.
   (F-IMPLIES X Y)

     =

   (F-OR (F-NOT X) Y)
```

`(FORALL X A)` is an abbreviation of the expression representing $\neg(\exists x. \neg A)$.

```
Definition.
   (FORALL VAR EXP)

     =

   (F-NOT (FORSOME VAR (F-NOT EXP)))
```

The function `PAIREQUALS` is used to construct the equality axioms. For example, if `PAIREQUALS` is given two lists of variables consisting of the x_i and the y_i, respectively, and a formula A, it returns a z-expression representing

$$((y_1 = z_1) \supset (\ldots \supset ((y_n = z_n) \supset A))).$$

```
Definition.
  (PAIREQUALS VARS1 VARS2 EXP)
    =
  (IF (AND (LISTP VARS1) (LISTP VARS2))
      (F-IMPLIES (F-EQUAL (CAR VARS1) (CAR VARS2))
                 (PAIREQUALS (CDR VARS1)
                             (CDR VARS2)
                             EXP))
      EXP)
```

The functions SECOND, THIRD, and FOURTH return the second, third and fourth elements of a list, respectively. The function CAR returns the first element.

```
Definition.
  (SECOND X)
    =
  (CADR X)
```

```
Definition.
  (THIRD X)
    =
  (CADDR X)
```

```
Definition.
  (FOURTH X)
    =
  (CADDDR X)
```

2.1.4 The Logical Axioms

In order to check the Lisp representations of Z2 proofs, we first need to be able to recognize instances of the logical axioms of Z2. We treat any instance of the axioms listed in Figures 1.4 and 1.5 as an axiom, so that the predicates defined below are recognizers for axiom schemes. The function FOL-AXIOM-PROOF checks if the conclusion CONC of a proof is a first-order logic axiom. In doing so, it uses hints provided in the list of hints HINTS. Depending on the value of (CAR HINTS), FOL-AXIOM-PROOF checks if:

1. CONC is a *propositional axiom*, of the form $(\neg A) \vee A$, where A is a formula.

2. CONC is a *substitution axiom*, of the form $([a/x]A \supset (\exists x.\ A))$, where A is a formula, x is a variable, and a is a term which is free for x in A.

3. CONC is an *identity axiom*, of the form $(a = a)$, where a is a term.

4. CONC is an *equality axiom for functions* of the form:

$$((a_1 = b_1) \supset (\ldots \supset ((a_n = b_n) \supset (f(a_1, \ldots, a_n) = f(b_1, \ldots, b_n))))),$$

where f is an n-ary function symbol and the a_i and b_i are terms.

5. CONC is an *equality axiom for predicates* of the form

$$((a_1 = b_1) \supset (\ldots \supset ((a_n = b_n) \supset (p(a_1, \ldots, a_n) \supset p(b_1, \ldots, b_n))))),$$

where p is an n-ary predicate symbol and the a_i and b_i are terms.

In the definition below, propositional axioms are checked to be of the form (F-OR (F-NOT A) A), where A is (SECOND HINTS). The identity axiom is also directly checked to be of the form (F-EQUAL A A), where A is (SECOND HINTS). The task of checking the remaining axioms is relegated to the functions SUB-AXIOM-PROOF, FUN-EQUALITY-AXIOM-PROOF, and PRED-EQUALITY-AXIOM-PROOF whose definitions follow.

```
Definition.
  (FOL-AXIOM-PROOF CONC HINTS SYMBOLS)
  =
  (IF (EQUAL (CAR HINTS) 1)
      (AND (FORMULA (SECOND HINTS) SYMBOLS)
           (EQL CONC
                (F-OR (F-NOT (SECOND HINTS))
                      (SECOND HINTS))))
      (IF (EQUAL (CAR HINTS) 2)
          (SUB-AXIOM-PROOF CONC HINTS SYMBOLS)
          (IF (EQUAL (CAR HINTS) 3)
              (AND (TERMP (SECOND HINTS) 0 SYMBOLS)
                   (EQL CONC
                        (F-EQUAL (SECOND HINTS)
                                 (SECOND HINTS))))
              (IF (EQUAL (CAR HINTS) 4)
                  (FUN-EQUALITY-AXIOM-PROOF CONC HINTS SYMBOLS)
                  (IF (EQUAL (CAR HINTS) 5)
                      (PRED-EQUALITY-AXIOM-PROOF CONC HINTS
                                                 SYMBOLS)
                      F)))))
```

The definition of SUB-AXIOM-PROOF checks that its CONC is of the form $[t/x]A \supset (\exists x.\ A)$, where A is (SECOND HINTS), x is (THIRD HINTS), and t is (FOURTH HINTS).

```
Definition.
   (SUB-AXIOM-PROOF CONC HINTS SYMBOLS)
     =
   (AND
     (FORMULA (SECOND HINTS) SYMBOLS)
     (AND (VARIABLE (THIRD HINTS))
          (AND (TERMP (FOURTH HINTS) 0 SYMBOLS)
               (AND (FREE-FOR (SECOND HINTS)
                              (THIRD HINTS)
                              (FOURTH HINTS)
                              0)
                    (EQL CONC
                         (F-IMPLIES
                          (SUBST (SECOND HINTS)
                                 (THIRD HINTS)
                                 (FOURTH HINTS)
                                 0)
                          (FORSOME (THIRD HINTS)
                                   (SECOND HINTS)))))))))
```

The definition of FUN-EQUALITY-AXIOM-PROOF uses PAIREQUALS to check that CONC has the form

$$((a_1 = b_1) \supset (\dots \supset ((a_n = b_n) \supset (f(a_1, \dots, a_n) = f(b_1, \dots, b_n))))),$$

where (SECOND HINTS) is a function symbol f, (THIRD HINTS) is a list of terms of the form $(a_1 \dots a_n)$, and (FOURTH HINTS) is of the form $(b_1 \dots b_n)$.

```
Definition.
   (FUN-EQUALITY-AXIOM-PROOF CONC HINTS SYMBOLS)
     =
   (AND
    (TERMP (THIRD HINTS) 1 SYMBOLS)
    (AND
     (TERMP (FOURTH HINTS) 1 SYMBOLS)
     (AND
      (FUNCTION (SECOND HINTS))
      (AND
       (EQUAL (LENGTH (THIRD HINTS))
              (DEGREE (SECOND HINTS)))
       (AND
        (EQUAL (LENGTH (FOURTH HINTS))
               (DEGREE (SECOND HINTS)))
        (EQL CONC
             (PAIREQUALS (THIRD HINTS)
                         (FOURTH HINTS)
                         (F-EQUAL (CONS (SECOND HINTS)
                                        (THIRD HINTS))
                                  (CONS (SECOND HINTS)
                                        (FOURTH HINTS)))))))))))
```

The definition of PRED-EQUALITY-AXIOM-PROOF is similar to that of FUN-EQUALITY-AXIOM-PROOF.

```
Definition.
  (PRED-EQUALITY-AXIOM-PROOF CONC HINTS SYMBOLS)

    =

  (AND
   (TERMP (THIRD HINTS) 1 SYMBOLS)
   (AND
    (TERMP (FOURTH HINTS) 1 SYMBOLS)
    (AND
     (PREDICATE (SECOND HINTS))
     (AND
      (EQUAL (LENGTH (THIRD HINTS))
             (DEGREE (SECOND HINTS)))
      (AND
       (EQUAL (LENGTH (FOURTH HINTS))
              (DEGREE (SECOND HINTS)))
       (EQL CONC
        (PAIREQUALS
                    (THIRD HINTS)
                    (FOURTH HINTS)
                    (F-IMPLIES (CONS (SECOND HINTS)
                                     (THIRD HINTS))
                               (CONS (SECOND HINTS)
                                     (FOURTH HINTS)))))))))))
```

This completes the formalization of the logical axioms of Z2 within the Lisp metatheory. We next consider the rules of inference followed by the nonlogical axioms.

2.1.5 The Rules of Inference

The function FOL-RULES checks if the conclusion CONC follows from the subgoal formulas in SUB-GOALS by an application of one of the rules of inference of first-order logic of Z2. The structure of FOL-RULES is similar to that of FOL-AXIOM-PROOF. Depending on the value of (CAR HINTS), FOL-RULES checks if:

1. CONC is of the form $A \lor B$, where B is the first subgoal.[4] This is the *Expansion* or *Weakening* rule.

2. CONC is of the form A, where $A \lor A$ is the only subgoal. This is the *Contraction* rule.

3. CONC is of the form $(A \lor B) \lor C$, where $A \lor (B \lor C)$ is the only subgoal. This is the *Associativity* rule.

4. CONC is of the form $B \lor C$, where the two subgoals are $A \lor B$ and $\neg A \lor C$. This is the *Cut* rule.

5. CONC is of the form $(\exists x.\ A) \supset B$, where the only subgoal is $A \supset B$, and x does not occur free in B. This is the $\exists$-*introduction* rule.

Depending on the value of (CAR HINTS), the function FOL-RULES invokes one of the functions EXPAND-RULE, CONTRACTION-RULE, ASSOC-RULE, CUT-RULE, or EXISTS-INTRO-RULE to check whether CONC follows from SUB-GOALS by an application of the relevant rule of inference of Z2.

---

[4]It would have been cleaner to check that this was the only subgoal, but no harm results.

```
Definition.
   (FOL-RULES CONC HINTS SUB-GOALS SYMBOLS)
      =
   (IF (EQUAL (CAR HINTS) 1)
       (EXPAND-RULE CONC HINTS SUB-GOALS SYMBOLS)
       (IF (EQUAL (CAR HINTS) 2)
           (CONTRACTION-RULE CONC HINTS SUB-GOALS SYMBOLS)
           (IF (EQUAL (CAR HINTS) 3)
               (ASSOC-RULE CONC HINTS SUB-GOALS SYMBOLS)
               (IF (EQUAL (CAR HINTS) 4)
                   (CUT-RULE CONC HINTS SUB-GOALS SYMBOLS)
                   (IF (EQUAL (CAR HINTS) 5)
                       (EXISTS-INTRO-RULE CONC HINTS SUB-GOALS
                                          SYMBOLS)
                       F)))))
```

The EXPAND-RULE function checks that the conclusion CONC is of the form (F-OR A B), where A is (SECOND HINTS) and B is the first subgoal formula.

```
Definition.
   (EXPAND-RULE CONC HINTS SUB-GOALS SYMBOLS)
      =
   (AND (FORMULA (SECOND HINTS) SYMBOLS)
        (AND (FORMULA (CAR SUB-GOALS) SYMBOLS)
             (EQL CONC
                  (F-OR (SECOND HINTS)
                        (CAR SUB-GOALS)))))
```

The CONTRACTION-RULE function checks if the conclusion CONC has the form A, where A is (SECOND HINTS), and the first subgoal has the form (F-OR A A).

```
Definition.
   (CONTRACTION-RULE CONC HINTS SUB-GOALS SYMBOLS)
      =
   (AND (FORMULA (SECOND HINTS) SYMBOLS)
        (AND (EQL SUB-GOALS
                  (CONS (F-OR (SECOND HINTS) (SECOND HINTS))
                        NIL))
             (EQL CONC (SECOND HINTS))))
```

The ASSOC-RULE function checks that the conclusion CONC has the form (F-OR (F-OR A B) C), where the first subgoal is of the form (F-OR A (F-OR B C)), and A, B, and C are the second, third, and fourth elements of the list HINTS, respectively.

```
Definition.
  (ASSOC-RULE CONC HINTS SUB-GOALS SYMBOLS)

      =

(AND
  (FORMULA (SECOND HINTS) SYMBOLS)
  (AND
   (FORMULA (THIRD HINTS) SYMBOLS)
   (AND
      (FORMULA (FOURTH HINTS) SYMBOLS)
      (AND (EQL SUB-GOALS
                (CONS (F-OR (SECOND HINTS)
                            (F-OR (THIRD HINTS)
                                  (FOURTH HINTS)))
                  NIL))
           (EQL CONC
                (F-OR (F-OR (SECOND HINTS) (THIRD HINTS))
                      (FOURTH HINTS)))))))
```

The CUT-RULE function checks that the conclusion CONC has the form (F-OR B C), where the two subgoals have the forms (F-OR A B) and (F-OR (F-NOT A) C), respectively, and A, B, and C are the second, third, and fourth elements of the list HINTS, respectively.

```
Definition.
  (CUT-RULE CONC HINTS SUB-GOALS SYMBOLS)

      =

(AND
  (FORMULA (SECOND HINTS) SYMBOLS)
  (AND (FORMULA (THIRD HINTS) SYMBOLS)
       (AND (FORMULA (FOURTH HINTS) SYMBOLS)
            (AND (EQL SUB-GOALS
                      (CONS (F-OR (SECOND HINTS)
                                  (THIRD HINTS))
                            (CONS (F-OR (F-NOT (SECOND HINTS))
                                        (FOURTH HINTS))
                                  NIL)))
                 (EQL CONC
                      (F-OR (THIRD HINTS)
                            (FOURTH HINTS)))))))
```

The EXISTS-INTRO-RULE function checks that the conclusion CONC has the form $(\exists x.\ A) \supset B$, where the first subgoal has the form $A \supset B$, and x, A, and B are the second, third, and fourth elements of the list HINTS, respectively.

```
Definition.
  (EXISTS-INTRO-RULE CONC HINTS SUB-GOALS SYMBOLS)
    =
  (AND
   (VARIABLE (SECOND HINTS))
   (AND
    (FORMULA (THIRD HINTS) SYMBOLS)
    (AND
     (FORMULA (FOURTH HINTS) SYMBOLS)
     (AND
      (NOT (MEMB (SECOND HINTS)
                 (COLLECT-FREE (FOURTH HINTS) '0)))
      (AND (EQL SUB-GOALS
                (CONS (F-IMPLIES (THIRD HINTS)
                                 (FOURTH HINTS))
                      'NIL))
           (EQL CONC
                (F-IMPLIES (FORSOME (SECOND HINTS)
                                    (THIRD HINTS))
                           (FOURTH HINTS)))))))))
```

2.1.6   The Set-Theoretic Axioms

In order to check proofs containing nonlogical axioms, we need to recognize instances of the extensionality, null-set, pairing, union, and induction axioms. We first define some notation for the set-theoretic operations of forming the null set, the pair set, the union set, the successor set, and also the element-of relation and is-an-integer predicate. The predicate symbol with index 1 and arity 2 is used to represent the set membership predicate symbol $\in$. (ISIN X Y) constructs the expression representing the atomic formula $(x \in y)$.

```
Definition.
  (ISIN X Y)
    =
  (LIST (P 1 2) X Y)
```

The expression representing the null set $\emptyset$ is (PHI). It is constructed by applying the function symbol of index 0 and arity 0 to no arguments.

```
Definition.
  (PHI)
    =
  (LIST (FN 0 0))
```

(Z-PAIR X Y) represents the term $\{x, y\}$, the pair set containing x and y.

```
Definition.
  (Z-PAIR X Y)
    =
  (LIST (FN 1 2) X Y)
```

(Z-UNION X Y) represents the term $(x \cup y)$, the union set of x and y.

```
Definition.
  (Z-UNION X Y)
    =
  (LIST (FN 2 2) X Y)
```

(Z-INT X) represents the atomic formula $I(x)$ which is defined to check if x is an ordinal.

```
Definition.
  (Z-INT X)
    =
  (LIST (P 2 1) X)
```

(Z-SUCC X) represents the term $S(x)$, the successor numeral of x.

```
Definition.
  (Z-SUCC X)
    =
  (LIST (FN 3 1) X)
```

(F-IFF A B) represents $(A \iff B)$ which is an abbreviation for the formula $(A \supset B) \wedge (B \supset A)$.

```
Definition.
  (F-IFF X Y)
    =
  (F-AND (F-IMPLIES X Y)
         (F-IMPLIES Y X))
```

The function Z2-AXIOMS checks if the given conclusion CONC is an instance of a nonlogical axiom of Cohen's Z2. According to the value of (CAR HINTS), Z2-AXIOMS uses hints from HINTS to check if:

1. CONC is an *Extensionality* axiom of the form:

$$a = b \iff (\forall x.\ x \in a \iff x \in b),$$

   where x does not occur free in a or in b.

2. CONC is a *Null set* axiom of the form:

$$\neg(a \in \emptyset).$$

3. CONC is a *Pairing* axiom of the form:

$$c \in \{a, b\} \iff c = a \vee c = b.$$

4. CONC is a *Union* axiom of the form:

$$c \in a \cup b \iff c \in a \vee c \in b.$$

5. CONC is an *Induction* axiom of the form:

$$([\emptyset/x]A \wedge (\forall x. \, ((I(x) \wedge A) \supset [S(x)/x]A))) \supset (\forall x. \, I(x) \supset A).$$

Here we do not check if $\emptyset$ and $S(x)$ are free for x in A since this is obviously the case. In this axiom, the symbol S is a defined function symbol, and I is a defined predicate symbol.[5] $S(x)$ is defined as $(\{x\} \cup x)$, where $\{x\}$ is the *singleton set*, $\{x, x\}$. $I(x)$ checks that x is transitive and linearly ordered by $\in$.

Z2-AXIOMS uses Z2-EXTENSIONALITY, Z2-NULLSET, Z2-PAIRING, Z2-UNION, and Z2-INDUCTION, to check the given conclusion. These definitions are given below.

```
Definition.
  (Z2-AXIOMS CONC HINTS SYMBOLS)
    =
  (IF (EQUAL (CAR HINTS) 1)
     (Z2-EXTENSIONALITY CONC HINTS SYMBOLS)
     (IF (EQUAL (CAR HINTS) 2)
        (Z2-NULLSET CONC HINTS SYMBOLS)
        (IF (EQUAL (CAR HINTS) 3)
           (Z2-PAIRING CONC HINTS SYMBOLS)
           (IF (EQUAL (CAR HINTS) 4)
              (Z2-UNION CONC HINTS SYMBOLS)
              (IF (EQUAL (CAR HINTS) 5)
                 (Z2-INDUCTION CONC HINTS SYMBOLS)
                 F)))))
```

Z2-EXTENSIONALITY checks if the conclusion has the form

$$a = b \iff (\forall x. \, x \in a \iff x \in b),$$

where x does not occur free in a or in b.

```
Definition.
  (Z2-EXTENSIONALITY CONC HINTS SYMBOLS)
    =
  (AND
   (VARIABLE (FOURTH HINTS))
   (AND
    (TERMP (SECOND HINTS) 0 SYMBOLS)
    (AND
     (TERMP (THIRD HINTS) 0 SYMBOLS)
     (AND
      (NOT (MEMB (FOURTH HINTS)
                 (COLLECT-FREE (SECOND HINTS) 0)))
      (AND
       (NOT (MEMB (FOURTH HINTS)
                  (COLLECT-FREE (THIRD HINTS) 0)))
       (EQL CONC
        (F-IFF
              (F-EQUAL (SECOND HINTS) (THIRD HINTS))
              (FORALL (FOURTH HINTS)
                  (F-IFF (ISIN (FOURTH HINTS)
                              (SECOND HINTS))
                       (ISIN (FOURTH HINTS)
                            (THIRD HINTS)))))))))))
```

---

[5]The definition of I is introduced in Section 2.1.7 along with the definitions of transitivity and linearity.

Z2-NULLSET checks if the conclusion has the form $\neg(a \in \emptyset)$.

```
Definition.
   (Z2-NULLSET CONC HINTS SYMBOLS)
    =
   (AND (TERMP (SECOND HINTS) 0 SYMBOLS)
        (EQL CONC
             (F-NOT (ISIN (SECOND HINTS) (PHI)))))
```

Z2-PAIRING checks if the conclusion has the form

$$(c \in \{a, b\}) \iff ((c = a) \lor (c = b)).$$

```
Definition.
   (Z2-PAIRING CONC HINTS SYMBOLS)
    =
   (AND (TERMP (CONS (SECOND HINTS)
                     (CONS (THIRD HINTS)
                           (CONS (FOURTH HINTS) NIL)))
              1
              SYMBOLS)
        (EQL CONC
             (F-IFF (ISIN (FOURTH HINTS)
                          (Z-PAIR (SECOND HINTS) (THIRD HINTS)))
                    (F-OR (F-EQUAL (FOURTH HINTS)
                                   (SECOND HINTS))
                          (F-EQUAL (FOURTH HINTS)
                                   (THIRD HINTS)))))))
```

Z2-UNION checks if the conclusion has the form

$$(c \in (a \cup b)) \iff ((c \in a) \lor (c \in b)).$$

```
Definition.
   (Z2-UNION CONC HINTS SYMBOLS)
    =
   (AND (TERMP (CONS (SECOND HINTS)
                     (CONS (THIRD HINTS)
                           (CONS (FOURTH HINTS) NIL)))
              1
              SYMBOLS)
        (EQL CONC
             (F-IFF (ISIN (FOURTH HINTS)
                          (Z-UNION (SECOND HINTS)
                                   (THIRD HINTS)))
                    (F-OR (ISIN (FOURTH HINTS) (SECOND HINTS))
                          (ISIN (FOURTH HINTS)
                                (THIRD HINTS)))))))
```

Z2-INDUCTION checks if the conclusion has the form

$$([\emptyset/x]A \land (\forall x.\ ((I(x) \land A) \supset [S(x)/x]A))) \supset (\forall x.\ I(x) \supset A).$$

```
Definition.
  (Z2-INDUCTION CONC HINTS SYMBOLS)
  =
(AND
  (VARIABLE (THIRD HINTS))
  (AND
  (FORMULA (SECOND HINTS) SYMBOLS)
  (EQL CONC
    (F-IMPLIES
        (F-AND (SUBST (SECOND HINTS)
                      (THIRD HINTS)
                      (PHI)
                      0)
              (FORALL (THIRD HINTS)
                   (F-IMPLIES (F-AND (Z-INT (THIRD HINTS))
                                     (SECOND HINTS))
                             (SUBST (SECOND HINTS)
                                    (THIRD HINTS)
                                    (Z-SUCC
                                     (THIRD HINTS))
                                    0))))
             (FORALL (THIRD HINTS)
                 (F-IMPLIES (Z-INT (THIRD HINTS))
                            (SECOND HINTS))))))))
```

2.1.7   Admitting Definitions

One difference between Shoenfield's first-order logic and the one formalized here is that we permit the introduction of axioms defining new predicate or function symbols. An n-ary predicate symbol p can be introduced by a defining axiom of the form $p(x_1 \ldots, x_n) \iff A$, where A contains only previously defined function and predicate symbols. An n-ary function symbol can be introduced with the defining axiom $[f(x_1, \ldots, x_n)/x]A$ provided the formula $(\exists!x.\ A)$ ("there exists a unique x such that A") is provable in the system without the function symbol f.

The main reason why the above extension is needed is that it is convenient to have function symbols for certain operations. This makes it easy to compose these operations with one another. For example, it is easier to say $(z = S(S(x)))$, than $(\exists y.\ (Is\text{-}successor(z, y) \wedge Is\text{-}successor(y, x)))$. The justification for this extension is given by a metatheorem which asserts that no new inconsistency is introduced into the logic by admitting such definitions as axioms. This metatheorem has an extremely complicated proof (see Kleene [Kle52, pages 407–415]). Such a metatheorem is not proved here but is cited as the informal justification for the use of definitional extensions. The model-theoretic justification of definitional extensions is simpler but well outside the scope of the present work.

(SET X) checks that no member of the list X has more than one occurrence in X. This is used to check if a given list of variables used in a definition is distinct.

```
Definition.
  (SET X)

    =

  (IF (LISTP X)
      (AND (NOT (MEMB (CAR X) (CDR X)))
           (SET (CDR X)))
      T)
```

(ASSOC X Y) returns a Z such that (CONS X Z) is the first pair in Y that has X as its first element. Z is then the value associated with X in the association-list Y. If no pair in Y is of the form (CONS X Z), then ASSOC returns 0. ASSOC is used to retrieve the definition corresponding to a function or predicate symbol from an association-list of symbols and definitions. Note that ASSOC is different from the standard Lisp definition of ASSOC which returns the pair. The non-LISTP elements of Y behave as (CONS 0 0).

```
Definition.
  (ASSOC X Y)

    =

  (IF (LISTP Y)
      (IF (EQL (CAAR Y) X)
          (CDAR Y)
          (ASSOC X (CDR Y)))
      0)
```

(REST-OF SYM SYMBOLS) returns the list that follows the last occurrence of SYM in the list SYMBOLS. It is used to ensure that there is no circularity in the definitions of the function and predicate symbols.

```
Definition.
  (REST-OF SYM SYMBOLS)

    =

  (IF (LISTP SYMBOLS)
      (IF (AND (EQL (CAR SYMBOLS) SYM)
               (NOT (MEMB SYM (CDR SYMBOLS))))
          (CDR SYMBOLS)
          (REST-OF SYM (CDR SYMBOLS)))
      NIL)
```

SYMB-DEFN-PROOF checks if CONC is admissible as an axiom defining a new function or predicate symbol whose definition is DEFN, given that the formulas in SUB-GOALS are provable within the theory whose symbols are from (REST-OF SYM SYMBOLS), where SYM is the symbol being defined. The first hint has to be of the form $p(x_1, \ldots, x_n)$ or $f(x_1, \ldots, x_n)$, depending on whether it is an n-ary predicate or a function symbol that is being defined.

If an n-ary predicate symbol is being defined, then the x_i must be distinct and CONC must be of the form $(p(x_1, \ldots, x_n) \iff A)$, where A is the DEFN and is a formula (with respect to the symbols that follow p in SYMBOLS) which contains no free variables other than the x_i.

If f is the n-ary function symbol that is being defined, then each of the following three conditions must hold:

1. The variable in the second hint x, the variable in the fourth hint y, and all the x_i must be distinct variables.

2. CONC must be the same as DEFN, and should be of the form $[f(x_1, \ldots, x_n)/x]A$, where A, the third hint, is a formula (with respect to the symbols that follow f in SYMBOLS) which contains no free variables other than x and the x_i. It is still possible that f could be defined in terms of itself. In order to rule this out, the definition of the proof checker PRF (Section 2.1.8) ensures that the formulas in SUB-GOALS are proved with respect to the symbols that follow f in SYMBOLS (i.e., the previously defined symbols).

3. SUB-GOALS must be a list of formulas that contains the *existence condition*, $(\exists x.\ A)$; and the *uniqueness condition*, $((A \wedge [y/x]A) \supset (x = y))$.

The checks in SYMB-DEFN-PROOF do not rule out the use of function or predicate symbols which are neither axiomatized nor defined. Since these could serve as free function variables, it is important for the sake of this proof to ensure that no such symbols are used. In Section 2.1.10 we show that every symbol in (THM-SYMBS) is either axiomatized, or is defined in (THM-DEFNS). The checks in SYMB-DEFN-PROOF do attempt to rule out the possibility of an inconsistency or an ambiguous interpretation.

```
Definition.
  (SYMB-DEFN-PROOF CONC HINTS SUB-GOALS DEFN SYMBOLS)
    =
  (IF (PREDICATE (CAAR HINTS))
      (AND (SET (CDAR HINTS))
           (VAR-LIST (CDAR HINTS)
                     (DEGREE (CAAR HINTS)))
           (EQL CONC (F-IFF (CAR HINTS) DEFN))
           (SUB (COLLECT-FREE DEFN 0)
                (CDAR HINTS))
           (FORMULA DEFN
                    (REST-OF (CAAR HINTS) SYMBOLS)))
      (IF (FUNCTION (CAAR HINTS))
          (AND (SET (CONS (FOURTH HINTS)
                          (CONS (SECOND HINTS) (CDAR HINTS))))
               (VAR-LIST (CDAR HINTS)
                         (DEGREE (CAAR HINTS)))
               (VARIABLE (SECOND HINTS))
               (VARIABLE (FOURTH HINTS))
               (FORMULA (THIRD HINTS)
                        (REST-OF (CAAR HINTS) SYMBOLS))
               (SUB (COLLECT-FREE (THIRD HINTS) 0)
                    (CONS (SECOND HINTS) (CDAR HINTS)))
               (FREE-FOR (THIRD HINTS)
                         (SECOND HINTS)
                         (CAR HINTS)
                         0)
               (FREE-FOR (THIRD HINTS)
                         (SECOND HINTS)
                         (FOURTH HINTS)
                         0)
               (EQL CONC
```

```
                (SUBST (THIRD HINTS)
                       (SECOND HINTS)
                       (CAR HINTS)
                       0))
        (EQL CONC DEFN)
        (EQL SUB-GOALS
                (LIST (FORSOME (SECOND HINTS) (THIRD HINTS))
                      (F-IMPLIES (F-AND (THIRD HINTS)
                                  (SUBST (THIRD HINTS)
                                         (SECOND HINTS)
                                         (FOURTH HINTS)
                                         0))
                             (F-EQUAL (SECOND HINTS)
                                      (FOURTH HINTS))))))))

    F))
```

2.1.8   The Proof Checker

An object-theoretic *proof* is represented as a Lisp data structure:

$$(\text{LIST } step \quad hints \quad conclusion \quad sub\text{-}proofs),$$

where *conclusion* is the conclusion of the proof, derived from the conclusions of the *sub-proofs* by a proof step indicated by the *step* component and the *hints* component.

(CONC PF FLG) returns the conclusion (list of conclusions) of a proof (list of proofs).

```
Definition.
  (CONC PF FLG)
      =
  (IF (ZEROP FLG)
      (THIRD PF)
      (IF (LISTP PF)
          (CONS (CONC (CAR PF) 0)
                (CONC (CDR PF) 1))
          NIL))
```

The predicate PRF checks if PF is a proof according to the rules and axioms of first-order logic, the axioms of Z2, and the admissibility rules for function and predicate definitions. GIVEN is a list of formulas that may be used as assumptions in the proof, DEFNS is the list of function and predicate definitions, and SYMBOLS is the list of permissible function and predicate symbols.

If FLG is 0, PRF checks if the conclusion of the proof is a formula with respect to SYMBOLS, and then according to the value of the step field of the 4-tuple PF, it checks whether the conclusion of the proof is:

1. A first-order logic axiom.

2. A consequence of the conclusions of the *sub-proofs* of PF by an application of one of the first-order logic rules, whereupon the sub-proofs themselves are checked to be well-formed proofs.

3. A Z2 axiom.

4. An admissible definition of a new function or predicate symbol, where (ASSOC (CAAR (SECOND PF)) DEFNS) is the appropriate definition from the list of definitions DEFNS. In the case when a function symbol is being defined, the existence and uniqueness conditions have to be proved by the sub-proofs of PF in a theory whose function and predicate symbols follow the last occurrence in SYMBOLS of the symbol being defined.

5. A formulas that is given as an assumption in GIVEN. This case is never used to actually construct any proofs in the theory, but is there to strengthen the proof of incompleteness so that we can show that the incompleteness theorem still holds even if additional axioms are added to Z2.

If FLG is 1, PRF checks if each member of the list PF is a well-formed proof with respect to GIVEN, DEFNS, and SYMBOLS. The definition is by recursion on the structure of the given proof PF.

```
Definition.
  (PRF PF GIVEN DEFNS FLG SYMBOLS)
     =
 (IF
  (ZEROP FLG)
  (IF
   (NLISTP PF)
   F
   (IF
     (NOT (FORMULA (CONC PF 0) SYMBOLS))
     F
     (IF (EQUAL (CAR PF) 1)
         (FOL-AXIOM-PROOF (CONC PF 0)
                          (SECOND PF)
                          SYMBOLS)
       (IF (EQUAL (CAR PF) 2)
           (AND (FOL-RULES (CONC PF 0)
                           (SECOND PF)
                           (CONC (FOURTH PF) 1)
                           SYMBOLS)
             (PRF (FOURTH PF)
                  GIVEN DEFNS 1 SYMBOLS))
         (IF (EQUAL (CAR PF) 3)
             (Z2-AXIOMS (CONC PF 0)
                        (SECOND PF)
                        SYMBOLS)
           (IF (EQUAL (CAR PF) 4)
               (AND (SYMB-DEFN-PROOF (CONC PF 0)
                                     (SECOND PF)
                                     (CONC (FOURTH PF) 1)
                                     (ASSOC
                                       (CAAR (SECOND PF))
```

```
                                      DEFNS)
                                   SYMBOLS)
                   (IF (FUNCTION (CAAR (CADR PF)))
                       (PRF (FOURTH PF)
                              GIVEN DEFNS 1
                              (REST-OF (CAAR (CADR PF))
                                       SYMBOLS))
                       T))
                  (IF (EQUAL (CAR PF) 5)
                      (MEMB (CONC PF 0) GIVEN)
                      F)))))))
        (IF (NLISTP PF)
            T
            (AND (PRF (CAR PF) GIVEN DEFNS 0 SYMBOLS)
                 (PRF (CDR PF)
                      GIVEN DEFNS 1 SYMBOLS))))
```

PROVES uses PRF to check that PF is a proof in the theory whose function and predicate symbols are in SYMBOLS, with assumptions from GIVEN, function and predicate definitions from DEFNS, and whose conclusion is the formula EXP. The predicate PROVES is a crucial part of the statement of the incompleteness theorem.

```
Definition.
   (PROVES PF EXP GIVEN DEFNS SYMBOLS)
   =
   (AND (EQL (CONC PF 0) EXP)
        (FORMULA EXP SYMBOLS)
        (PRF PF GIVEN DEFNS 0 SYMBOLS))
```

The above formalization of the logic differs from those of Shoenfield [Sho67] and Cohen [Coh66] in the following ways:

1. All instances of axioms are also recognized as axioms so that each axiom is treated as an axiom scheme.

2. Shoenfield proves a metatheorem which demonstrates that the addition of function and predicate definitions which satisfy the requirements specified above, constitutes a *conservative extension*, that is, the theory with the new axioms is consistent if the original theory is consistent. We have instead built this into the definition of PRF above.

3. The induction axiom is a slightly simplified version of the one given by Cohen.[6]

2.1.9 An Example of a Formal Proof

A simple example of a formal proof in the above theory will be described below. It deals solely with the propositional part of the logic. We wish to construct a formal proof of the

---

[6]Induction only plays a minor role in the proof of the incompleteness theorem. It is used in only one proof, namely, that of Lemma 4.3.4 (see page 103).

statement $\neg A \vee (\neg B \vee A)$, where A represents the formula $(x_0 = x_1)$ and B represents $(x_3 = x_2)$. The formal proof of $\neg A \vee (\neg B \vee A)$ is first presented in its textual form. The Lisp data structure encoding this formal proof is then displayed. The data structure has been checked by PROVES as being a correct proof of the above statement.

$$
\cfrac{
  \cfrac{
    \cfrac{\overline{\neg A \vee A}^{\ Axiom} \qquad \overline{(\neg\neg A) \vee \neg A}^{\ Axiom}}{A \vee \neg A}{}^{Cut}
  }{
    \cfrac{\neg B \vee (A \vee \neg A)}{(\neg B \vee A) \vee \neg A}{}^{Assoc}
  }{}^{Expansion}
  \qquad
  \overline{\neg(\neg B \vee A) \vee (\neg B \vee A)}^{\ Axiom}
}{
  \neg A \vee (\neg B \vee A)
}{}^{Cut}
$$

The Lisp data structure representing the above proof is shown below. The symbol '(3 0 . 2) represents the equality predicate. The symbols 0, 1, 2, and 3 are x_0, x_1, x_2, and x_3, respectively. So the form '((3 0 . 2) 0 1) represents $(x_0 = x_1)$. The top level of the proof has the form

    (LIST 2 (LIST 4 hint1 hint2 hint3) conclusion (LIST pf1 pf2)).

The first element 2 indicates that the last step is an application of one of the logical rules of inference. The 4 in the hints field of the proof indicates that the step is an application of the *Cut* rule. The sub-proofs are pf1 and pf2, where pf2 has the form

    (LIST 1 (LIST 1 hint1) conclusion).

The first 1 indicates that the conclusion of the sub-proof is a logical axiom, and the second 1 indicates that it is a propositional axiom.

```
(LIST
  2
  (LIST 4
      (F-OR (F-NOT '((3 0 . 2) 3 2))
            '((3 0 . 2) 0 1))
      (F-NOT '((3 0 . 2) 0 1))
      (F-OR (F-NOT '((3 0 . 2) 3 2))
            '((3 0 . 2) 0 1)))
  (F-OR (F-NOT '((3 0 . 2) 0 1))
      (F-OR (F-NOT '((3 0 . 2) 3 2))
            '((3 0 . 2) 0 1)))
  (LIST
    (LIST 2
        (LIST 3
            (F-NOT '((3 0 . 2) 3 2))
            '((3 0 . 2) 0 1)
            (F-NOT '((3 0 . 2) 0 1)))
        (F-OR (F-OR (F-NOT '((3 0 . 2) 3 2))
                    '((3 0 . 2) 0 1))
            (F-NOT '((3 0 . 2) 0 1)))
        (LIST
          (LIST 2
              (LIST 1 (F-NOT '((3 0 . 2) 3 2)))
              (F-OR (F-NOT '((3 0 . 2) 3 2))
                  (F-OR '((3 0 . 2) 0 1)
```

```
                            (F-NOT '((3 0 . 2) 0 1))))
              (LIST
                (LIST
                  2
                  (LIST 4
                        (F-NOT '((3 0 . 2) 0 1))
                        '((3 0 . 2) 0 1)
                        (F-NOT '((3 0 . 2) 0 1)))
                  (F-OR '((3 0 . 2) 0 1)
                        (F-NOT '((3 0 . 2) 0 1)))
                  (LIST
                    (LIST 1
                          '(1 ((3 0 . 2) 0 1))
                          (F-OR (F-NOT '((3 0 . 2) 0 1))
                                '((3 0 . 2) 0 1)))
                    (LIST 1
                          (LIST 1 (F-NOT '((3 0 . 2) 0 1)))
                          (F-OR
                            (F-NOT (F-NOT '((3 0 . 2) 0 1)))
                            (F-NOT '((3 0 . 2) 0 1)))))))))))
    (LIST 1
          (LIST 1
                (F-OR (F-NOT '((3 0 . 2) 3 2))
                      '((3 0 . 2) 0 1)))
          (F-OR (F-NOT (F-OR (F-NOT '((3 0 . 2) 3 2))
                             '((3 0 . 2) 0 1)))
                (F-OR (F-NOT '((3 0 . 2) 3 2))
                      '((3 0 . 2) 0 1)))))))
```

This concludes the presentation of a Lisp data structure representing a formal proof that has been checked by the proof checker PROVES. It should be clear that constructing proofs solely in terms of the axioms and rules of the formal theory is a fairly tedious process even for simple theorems. In the next chapter we show how it is possible to employ more convenient rules of inference to construct formal proofs in a sound manner.

2.1.10   The Function and Predicate Definitions

To complete the details of the formalization of the incompleteness theorem, we need to show that every symbol in (THM-SYMBS) is either axiomatized directly in Z2 or defined in (THM-DEFNS). The symbols in (THM-SYMBS) are:

```
    Definition.
      (THM-SYMBS)
        =
      (LIST (FN 16 1)(FN 15 2)(FN 14 2)(FN 13 2)(FN 12 2)
            (FN 11 2)(FN 10 1)(FN 9 1)(FN 8 1)(FN 7 1)(FN 6 1)
            (FN 5 1)(FN 4 1)(FN 3 1)(P 2 1)(FN 2 2)(FN 1 2)(P 1 2)
            (FN 0 0))
```

Of these, (FN 0 0) represents the Null set, (P 1 2) represents $\in$, (FN 1 2) represents Pairing, (FN 2 2) represents Union; all are axiomatized in Z2. The remaining symbols are matched to definitions in the association-list forming the body of THM-DEFNS below.

```
Definition.
   (THM-DEFNS)
      =
   (LIST (CONS (P 2 1) (Z-INT-DEFN 0 1 2))
          (CONS (FN 3 1)
                  (F-EQUAL (Z-SUCC 0)
                            (Z-UNION (Z-SING 0) 0)))
          (CONS (FN 4 1)
                  (Z-CAR-DEFN 0 (Z-CAR 0) 1 2))
          (CONS (FN 5 1)
                  (Z-CDR-DEFN 0 (Z-CDR 0) 1 2))
          (CONS (FN 6 1)
                  (Z-LISTP-DEFN 0 (Z-LISTP 0) 1 2))
          (CONS (FN 7 1)
                  (Z-ADD1-DEFN 0 (Z-ADD1 0) 1 2))
          (CONS (FN 8 1)
                  (Z-NUMBERP-DEFN 0 (Z-NUMBERP 0)))
          (CONS (FN 9 1)
                  (Z-ZEROP-DEFN 0 (Z-ZEROP 0)))
          (CONS (FN 10 1)
                  (Z-SUB1-DEFN 0 (Z-SUB1 0) 3))
          (CONS (FN 11 2)
                  (Z-EQUAL-DEFN 0 1 (Z-EQUAL 0 1)))
          (CONS (FN 12 2)
                  (Z-CONS-DEFN 0 1 (Z-CONS 0 1)))
          (CONS (FN 13 2)
                  (Z-APPLY-SUBR1-DEFN 0 1
                                        (Z-APPLY-SUBR1 0 1)))
          (CONS (FN 14 2)
                  (Z-APPLY-SUBR2-DEFN 0 1
                                        (Z-APPLY-SUBR2 0 1)))
          (CONS (FN 15 2)
                  (Z-APPLY-SUBR-DEFN 0 1
                                        (Z-APPLY-SUBR 0 1)))
          (CONS (FN 16 1)
                  (Z-SUBRP-DEFN 0 (Z-SUBRP 0))))
```

Of these symbols, (P 2 1) represents the predicate symbol I, and (FN 3 1) represents the function symbol S. Since these symbols are used in the Induction axiom, their definitions are important. The definition associated with S in (THM-DEFNS) is

```
(F-EQUAL (Z-SUCC 0)
          (Z-UNION (Z-SING 0) 0)).
```

(Z-SING X) represents the *singleton set* $\{x\}$. Thus the defining axiom of S can be read as $S(x) = (\{x\} \cup x)$. (Note that 0 in the definition above stands for the first formal variable and not the numeral zero.)

```
Definition.
   (Z-SING X)
      =
   (Z-PAIR X X)
```

The definition of the I predicate is given by the following series of definitions leading up to Z-INT-DEFN. They can be seen to define $I(x)$ as:

$$(\forall y.\ (\forall z.\ (((y \in x) \wedge (z \in y)) \supset (z \in x))))$$
$$\wedge\quad (\forall y.\ (\forall z.\ (((y \in x) \wedge (z \in x)) \supset ((y = z) \vee ((y \in z) \vee (z \in y)))))).$$

The first condition asserts that x is a transitive set, that is, any z that is an element of y in x, is also an element of x. The second asserts that x is linearly ordered by $\in$: for any given y and z in x, either y is equal to z or is an element of z, or z is an element of y.

```
Definition.
  (Z-TRICH1 X Y Z)

    =

  (F-IMPLIES (F-AND (ISIN Y X) (ISIN Z X))
             (F-OR (F-EQUAL Y Z)
                   (F-OR (ISIN Y Z) (ISIN Z Y))))
```

```
Definition.
  (Z-TRICH X V1 V2)

    =

  (FORALL V1
          (FORALL V2 (Z-TRICH1 X V1 V2)))
```

```
Definition.
  (Z-TRANS1 X U V)

    =

  (F-IMPLIES (F-AND (ISIN U X) (ISIN V U))
             (ISIN V X))
```

```
Definition.
  (Z-TRANS X Y Z)

    =

  (FORALL Y (FORALL Z (Z-TRANS1 X Y Z)))
```

```
Definition.
  (Z-INT-DEFN X Y Z)

    =

  (F-AND (Z-TRANS X Y Z)
         (Z-TRICH X Y Z))
```

The other definitions in (THM-SYMBS) are irrelevant to the correctness of the statement of the incompleteness theorem since we only need to ensure that no undefined or unaxiomatized symbols were used. These definitions are of course important to the proof of the incompleteness theorem and they are listed in Section 4.4.1. This completes the details of the formalized statement of the incompleteness theorem.

2.2   The Formal Statement of the Incompleteness Theorem

The conventional statement of the incompleteness theorem for Z2 asserts that the consistency of Z2 implies the existence of a sentence that is neither provable nor disprovable in Z2. Such a statement cannot be formalized in the Boyer–Moore logic since the consistency hypothesis requires the use of quantification. The statement of the incompleteness theorem formulated below is the contrapositive form of the above statement. It asserts the existence of a sentence (i.e., a formula with no free variables) in the logic, which if either provable or disprovable would be both provable and disprovable. Such a sentence is said to be *undecidable*. In simple terms, the theory Z2 is either incomplete or inconsistent. The statement of the incompleteness theorem will be described in this section.

The undecidable sentence is constructed by a function which takes GIVEN as an argument. The undecidable sentence is therefore represented by the term (UNDECIDABLE-SENTENCE GIVEN). The negation of the undecidable sentence is the term (F-NOT (UNDECIDABLE-SENTENCE GIVEN)). Thus we can construct an undecidable sentence for any assumption list GIVEN so that the addition of any particular undecidable sentence to the assumption list only yields a new undecidable sentence. The theory Z2 is therefore proved to be *essentially incomplete*.

The statement of the incompleteness theorem is the conjunction of three assertions about the undecidable sentence:

1. (UNDECIDABLE-SENTENCE GIVEN) is a well-formed formula with respect to a particular list of symbols (THM-SYMBS).

2. The undecidable sentence is indeed a sentence.

3. Given any PF which is either a proof or a disproof of the undecidable sentence with respect to the assumption-list GIVEN and the function and predicate definitions (THM-DEFNS), in a theory whose function and predicate symbols are among (THM-SYMBS), it is possible to construct both a proof and a disproof of the undecidable sentence. Since the Boyer–Moore logic does not provide existential quantification, the existence of the proof and disproof has to be asserted by defining functions SOME-PROOF and SOME-DISPROOF which explicitly construct them. Once these are shown to exist, the details of their actual construction become irrelevant to the correctness of the statement. The conclusion of the implication is a disjunction, where both disjuncts assert the existence of proofs and disproofs. These disjuncts correspond to the two cases that arise in the informal argument presented in Chapter 1.

The statement is given on the next page.

To completely justify the correctness of that statement of the incompleteness theorem, we only need examine the definitions of F-NOT, SENTENCE, FORMULA, and PROVES to make sure that they correctly capture the metatheory of Z2. We also need to ensure that all the symbols in (THM-SYMBS) are either axiomatized in PROVES or defined in (THM-DEFNS). The details of the construction of the undecidable sentence (UNDECIDABLE-SENTENCE GIVEN) and the definitions of the functions SOME-PROOF1, SOME-PROOF2, SOME-DISPROOF1, and SOME-DISPROOF2 are not needed to achieve the conviction that the incompleteness theorem has been correctly formulated.

```
Theorem.   INCOMPLETENESS (rewrite):
   (AND
    (FORMULA (UNDECIDABLE-SENTENCE GIVEN)
             (THM-SYMBS))
    (AND
     (SENTENCE (UNDECIDABLE-SENTENCE GIVEN))
     (IMPLIES
             (OR (PROVES PF
                        (UNDECIDABLE-SENTENCE GIVEN)
                        GIVEN
                        (THM-DEFNS)
                        (THM-SYMBS))
                (PROVES PF
                        (F-NOT (UNDECIDABLE-SENTENCE GIVEN))
                        GIVEN
                        (THM-DEFNS)
                        (THM-SYMBS)))
             (OR (AND (PROVES (SOME-PROOF1 GIVEN PF)
                        (UNDECIDABLE-SENTENCE GIVEN)
                        GIVEN
                        (THM-DEFNS)
                        (THM-SYMBS))
                     (PROVES (SOME-DISPROOF1 GIVEN PF)
                        (F-NOT
                          (UNDECIDABLE-SENTENCE GIVEN))
                        GIVEN
                        (THM-DEFNS)
                        (THM-SYMBS)))
                (AND (PROVES (SOME-PROOF2 GIVEN PF)
                        (UNDECIDABLE-SENTENCE GIVEN)
                        GIVEN
                        (THM-DEFNS)
                        (THM-SYMBS))
                     (PROVES (SOME-DISPROOF2 GIVEN PF)
                        (F-NOT
                          (UNDECIDABLE-SENTENCE GIVEN))
                        GIVEN
                        (THM-DEFNS)
                        (THM-SYMBS)))))))
```

2.3   Summary

The formal statement of the incompleteness theorem for Z2 was presented in terms of a proof checker for Z2. The details of the construction of the proof checker were carefully examined. If the incompleteness theorem has been correctly formalized, then the correctness of the proof depends only on the soundness of the Boyer–Moore theorem prover. The formalization of the incompleteness theorem can be thoroughly scrutinized since the details only cover about twenty pages of pure Lisp.

The basic goal of the proof is to construct an undecidable sentence in the theory Z2, and to demonstrate its undecidability. This sentence will involve some assertion regarding its own provability or disprovability. The proof can be broken up into the following parts:

1. Proofs of metatheorems about the PROVES predicate which can be used as *derived inference rules.*

2. Development of a small amount of set theory, mainly to do with theorems about ordered pairs and finite ordinals.

3. Definition of a Lisp interpreter EV which is shown to be *representable* in Z2. EV formalizes the notion of computability used in the proof.

4. Proof of the computability of the metatheoretic functions such as PROVES, SUBST, etc.

5. Construction of the undecidable sentence and proof that it is undecidable with respect to Z2.

Step 1 will be described in the next chapter. Steps 2, 3 and 4 will be described in Chapter 4, and step 5 will be described in Chapter 5.

Chapter 3

Derived Inference Rules

> *But formalized mathematics cannot*
> *in practice be written down in full . . . .*
> *We shall therefore very quickly*
> *abandon formalized mathematics.*
> N. Bourbaki [Bou68]

In the previous chapter we defined a proof-checking program to check formal proofs constructed according to the axioms and rules of inference of the formal system Z2. In order to make progress towards the goal of proving the incompleteness of Z2, we need to formally develop a small amount of mathematics within Z2. While it is true that constructing formal proofs is very tedious, we have no intention of abandoning formalized mathematics. We will show in this chapter that formal proofs can be constructed in bigger and more natural steps by using powerful derived inference rules.[1] These are inference rules whose application can always be eliminated in favor of the primitive axioms and inference rules of Z2. In other words, the *soundness* of these rules can be proven as a metatheorem. These derived rules do not yield any new theorems but they make the process of formal proof construction more natural and less laborious.

We demonstrate how the Boyer–Moore theorem prover can be used to prove and use derived inference rules to construct nontrivial formal proofs. Most of this chapter is an exposition of one powerful derived inference rule which states that all propositional tautologies are theorems of Z2. We also list several other derived inference rules that were similarly proved.[2]

---

[1]Some of the results described in this chapter have been previously published in the *Journal of Automated Reasoning* [Sha85]. Copyright 1985, Kluwer Academic Publishers. Revised and reprinted by permission.

[2]A *derived* inference rule of a theory is a rule where there is a deduction in the theory of the conclusion of the rule from its premises. An *admissible* inference rule is one where the conclusion of the rule is provable whenever its premises are provable in the theory. Every derived inference rule is clearly admissible. The theorems presented in this chapter establish the admissibility of various inference rules that are in fact derivable. The mechanically verified soundness proofs for these rules do explicitly demonstrate the existence of a deduction of the conclusion of each rule from its premises.

3.1 Metamathematical Extensibility

While it is, in principle, possible to use the proof checker PROVES to construct and check formal proofs of interesting theorems, it would require an unrealistic amount of labor. If formal proof checking is to be feasible, it should be possible to extend the proof checker with inference rules that are closer to those employed in informal proof constructions. There is a danger that such extensions might introduce an unsoundness so that the extended proof checker proves a falsehood. It would be preferable if such extensions did not prove any theorems that the basic proof checker could not already prove.

The ability to make sound extensions to a formal proof checker has been termed *meta-mathematical extensibility* [Dav65]. There have been two approaches to building extensible proof checkers. The first approach uses the proof checker to prove the soundness of new extensions to itself. This approach has been used in FOL [WT74] and in the Boyer-Moore theorem prover [BM81]. The key idea in the FOL system is to start with a simple proof checker that incorporates the inference rules of first-order logic along with some nonlogical axioms. This proof checker is used to *reflexively* define its own metatheory so that one can prove metatheorems about the formulas, axioms, and proof rules used by the system within FOL itself. Any derived inference rules proved in this manner can be added to the system through a process of *semantic attachment* where program code implementing the derived inference rules is introduced, but this latter step is a potential source of human error.

In the second approach to extensibility, one avoids proving the soundness of new extensions by ensuring that every time a derived rule is used, the expanded formal proof corresponding to the relevant instance of this rule is generated and checked. This expanded proof can be automatically generated by executing a program which constructs the correct formal proof to justify the use of that proof-step. The expanded proof can then be checked using the original proof checker, but this process will be time-consuming since formal proofs tend to be large. This approach has been used in Edinburgh LCF [GMW79][3] and by Brown [Bro80]. The resulting derived inference procedures are called *tactics* in LCF.

In this chapter we actually prove several derived inference rules to be sound by demonstrating that their application can always be replaced by a proof involving only the primitive axioms and inference rules. Once we have proved this, we need never expand out the proof since it is sufficient to know that such a proof exists. Our approach is therefore quite similar to the FOL approach in this sense, but we do not prove these metatheorems by reflecting them into the object theory. We instead employ the Boyer–Moore theorem prover to prove the metatheorems since it is quite a powerful tool for this purpose. One drawback of our approach is that to believe in the soundness of the resulting derived inference rule, one must trust the soundness of the Boyer–Moore theorem prover. The soundness of the theorem prover is a serious issue, but one can derive some assurance from the fact that it is widely used and quite thoroughly tested.

The proofs presented in this chapter introduce derived inference rules extending the proof checker PROVES described in the previous chapter. The most significant such derived inference rule is the tautology theorem. This was the first significant result in metamathematics (proved by Post in his doctoral dissertation [Pos21], and independently by Bernays [Ber26]).

---

[3]It is also used in the many descendants of Edinburgh LCF including Cambridge LCF [Pau87], HOL [Gor88], Nuprl [Con86], and the Calculus of Constructions [CH85], among others.

A tautology is a propositional expression whose truth-table evaluation under any assignment of **true** or **false** to the atomic expressions always yields **true**, given the usual interpretation of the logical connectives. Our version of the tautology theorem states that every tautology has a proof in Shoenfield's first-order logic. This theorem justifies one of the most commonly used rules of inference in informal mathematical arguments. It also proves the propositional completeness of the formal logic.

Several other important derived inference rules have also been similarly proved. These include the properties of the various connectives and quantifiers, and rules regarding the use of equality, equivalence, and instantiation. Some of these rules will be discussed briefly, but the bulk of the presentation is devoted to the mechanical proof of the tautology theorem.

Section 3.2 is an outline of the informal proof of the tautology theorem. Section 3.3 is the longest section in the chapter and describes the mechanical proof of the tautology theorem. Section 3.4 lists some of the other important derived inference rules that were proved sound.

3.2   The Informal Proof of the Tautology Theorem

We will present the informal proof of the tautology theorem for Shoenfield's first-order logic. We need only pay attention to the propositional part of the logic and need not attach any specific interpretation to quantified formulas or atomic formulas. Therefore, these formulas will be treated as *propositional atoms* (atoms, for short). *Propositional formulas* are constructed by combining atoms using the operators $\neg$ and $\vee$. It should be clear that any Z2 formula can be construed as a propositional formula.

The proof consists of the following parts:

1. Definition of a tautology checker.

2. The proof of a useful lemma which states that from a proof of the disjunction of a list of formulas Γ, one can construct the proof of the disjunction of any larger list of formulas Δ containing all the formulas in Γ.

3. The proof of the tautology theorem.

3.2.1   A Tautology Checker

Logical truth is defined by the use of a truth table as we have already seen in page 8. Given a propositional formula, a *truth assignment* for that formula is a mapping from the set of propositional atoms to {**true, false**}. A *tautology* is defined as a propositional formula whose truth value is **true** under all truth assignments. A *tautology checker* is an algorithm that checks if a given propositional formula is a tautology or not. We shall define one such tautology checker. This tautology checker works only on formulae of the form $A_1 \vee \ldots \vee A_n$ where each A_i is termed a *disjunct*. Note that rearranging the disjuncts does not change the truth value, and that any formula A can be expressed in this form by simply setting A_1 to be A and $n = 1$.

The tautology checker $TC(A)$ is defined by recursion on the size of the formula A. In the base case, each A_i is either an atom or the negation of an atom, and there are two cases to the definition:

$$TC(A_1 \vee \ldots \vee A_n) = \textbf{true}, \text{ if for some } i, j \colon A_i \equiv (\neg A_j) \qquad (3.1)$$

$$TC(A_1 \vee \ldots \vee A_n) = \textbf{false}, \text{ otherwise.} \qquad (3.2)$$

Otherwise, some A_i is neither an atom nor the negation of an atom. Since the disjuncts can be rearranged without affecting the truth value, we can assume that A_1 is such an A_i. There are three cases according to whether A_1 is of the form $B \vee C$, the form $\neg(B \vee C)$, or the form $\neg\neg B$.

$$TC((B \vee C) \vee \ldots \vee A_n) = TC(B \vee C \vee \ldots \vee A_n) \qquad (3.3)$$

$$TC((\neg(B \vee C)) \vee \ldots \vee A_n) = \left\{ \begin{array}{c} TC((\neg B) \vee \ldots \vee A_n) \\ \text{and} \quad TC((\neg C) \vee \ldots \vee A_n) \end{array} \right. \qquad (3.4)$$

$$TC((\neg\neg B) \vee \ldots \vee A_n) = TC(B \vee \ldots \vee A_n) \qquad (3.5)$$

The sum of the number of logical operators in the A_i decreases with each recursive call and therefore the algorithm always terminates. To show that the above tautology checker is correct, we need to prove that $TC(A) = \textbf{true}$ if and only if A is a tautology. We can conclude that if $TC(A) = \textbf{true}$ then A is a tautology, by carrying out an inductive proof based on the recursion used in the tautology checker. Showing that if A is a tautology, then $TC(A) = \textbf{true}$ is a little more difficult. The proof involves following the same induction to construct a truth assignment that makes the truth value of A equal to **false** when $TC(A) = \textbf{false}$. The correctness of TC is stated as Theorem 3.2.1, and the details of its proof are left to the reader.

Theorem 3.2.1 (Correctness) *A is a tautology iff $TC(A) = \textbf{true}$.*

3.2.2 The Tautology Theorem

The proof of the tautology theorem demonstrates that every tautology has a proof in Z2. We first state a key 'lemma on disjuncts' that is extremely useful in the proof. The reader can consult pages 28–29 of Shoenfield [Sho67] for the proof. In the statements below and in the following chapters, '⊢ A' denotes 'A is a theorem of Z2.' The relationship between the PROVES predicate and the '⊢' symbol needs a bit of explanation. The PROVES predicate takes as its arguments the proof PF, the theorem being proved EXP, the assumed formulas GIVEN, the function and predicate definitions DEFNS, and the function and predicate symbols SYMBOLS. The argument GIVEN is included so that we can eventually establish that Z2 is essentially incomplete by constructing an undecidable sentence that is parametric in GIVEN. The argument GIVEN is therefore a fixed parameter for the entire exposition leading up to the proof of the incompleteness theorem. For a fixed GIVEN, the theory Z2 refers to one where the DEFNS argument is set to (THM-DEFNS) and the SYMBOLS argument is set to (THM-SYMBS). So if A is the informal representation of a Z2 formula represented in Lisp by A, then ⊢ A means that there exists a proof PF such that (PROVES PF A GIVEN (THM-DEFNS) (THM-SYMBS)) is T.

Lemma 3.2.2 *If for each i, $0 < i \leq m$, there is a j, $0 < j \leq n$, such that A_i is identical to B_j and ⊢ $A_1 \vee \ldots \vee A_m$, then ⊢ $B_1 \vee \ldots \vee B_n$.*

We now state the main theorem.

Theorem 3.2.3 (tautology theorem) *If A is a tautology, then $\vdash A$.*

Proof. We informally sketch the proof of the tautology theorem. The sketch should serve as a useful guide to the mechanical proof presented in the following section.

Since A is a tautology if and only if $(A \vee A)$ is a tautology, and we can derive A from $(A \vee A)$ by the Contraction rule, we can restrict our attention to formulas of the form $A_1 \vee \ldots \vee A_n$, where n is at least 2.

The proof of the tautology theorem is by an induction identical to the recursion displayed by the tautology checker TC. Since we have argued that the tautology checker TC as defined earlier is correct, we need only show that if the tautology checker accepts a formula A of the form $A_1 \vee \ldots \vee A_n$ (where n is at least 2) as being a tautology, then we can construct a proof of it in the propositional part of Z2. Let us assume that the given formula is a tautology; then the following cases arise:

Case 1: *All A_i are atoms or negations of atoms*: By the definition of the tautology checker, some A_j must be the negation of some A_i. Then, $\vdash A_j \vee A_i$ (Propositional Axiom) from which we get $\vdash A$ by applying Lemma 3.2.2.

Case 2: *Some A_i is of the form $B \vee C$*: By rearranging the disjuncts using Lemma 3.2.2, we can ensure that $i = 1$. We have $\vdash B \vee C \vee A_2 \vee \ldots \vee A_n$ by the induction hypothesis and the definition of the tautology checker. From this we derive $\vdash A$ by an application of the Associativity rule.

Case 3: *A_1 is of the form $\neg(B \vee C)$*: Again, by examining the tautology checker, we get $\vdash \neg B \vee A_2 \vee \ldots \vee A_n$ and $\vdash \neg C \vee A_2 \vee \ldots \vee A_n$, by the induction hypothesis. A proof of $\vdash A$ can be constructed from the proofs of these two formulas. The details are left to the reader.

Case 4: *A_1 is of the form $\neg\neg B$*: By the definition of the tautology checker and the induction hypothesis, we have $\vdash B \vee A_2 \vee \ldots \vee A_n$. The construction of the proof of $\vdash A$ from the proof of $\vdash B \vee A_2 \vee \ldots \vee A_n$ is left to the reader.

∎

The proof sketched above serves as a guide to the mechanical verification described below.

3.3   The Mechanical Proof of the Tautology Theorem

In this section, we cover some of the highlights of the mechanical proof. The proof can roughly be divided into the following parts:

1. Proof of the Lemma 3.2.2 on disjuncts.

2. Definition of the tautology checker.

3. Proof of the tautology theorem.

4. Proof of correctness of the tautology checker.

3.3.1   Steps to the Proof of the Tautology Theorem

This section lists some of the important lemmas involved in the mechanical proof of the tautology theorem. The form of the interaction with the theorem prover should also be noted since it can have a considerable impact on the success of a mechanical proof attempt. For example, immediately after the proof checker PRF is defined , we replace its definition by a series of lemmas that the theorem prover uses as rewrite rules. The definition of PRF needs to be disabled because it is long and causes a great deal of garbage generation and collection during the proof. In most of the lemmas that we prove, only a small part of the definition of the proof checker is relevant. We provide a one simple example of a rewrite rule that captures one case of the definition of PRF, the one dealing with propositional axioms. The other cases are similar. The function PROP-AXIOM-PROOF constructs a proof of a propositional axiom (as constructed by PROP-AXIOM) that can be checked by PRF.

```
Definition.
   (PROP-AXIOM EXP)
      =
   (F-OR (F-NOT EXP) EXP)
```

```
Definition.
   (PROP-AXIOM-PROOF EXP)
      =
   (CONS 1
         (CONS (CONS 1 (CONS EXP NIL))
               (CONS (PROP-AXIOM EXP) NIL)))
```

The lemma PROP-AXIOM-PROVES attempts to rewrite any term in a proof of the form

```
(PROVES (PROP-AXIOM-PROOF expression) conclusion given defns symbols)
```

to T, if expression is a formula and conclusion is a propositional axiom involving expression. Thus, if PROP-AXIOM-PROOF is used to construct a Z2 proof, this lemma will be invoked in the course of checking that proof. It is important to note that, once we have this lemma, the actual definition of PROP-AXIOM-PROOF is no longer useful. A definition to the Boyer–Moore theorem prover can be "disabled," that is, the prover cannot employ the definition of the disabled function in a proof until it has been "enabled." Following the theorem below, the definition of the *proof constructor* PROP-AXIOM-PROOF is disabled so that the data structure it represents is never generated in the process of checking a Z2 proof construction.

```
Theorem.  PROP-AXIOM-PROVES (rewrite):
   (IMPLIES (FORMULA EXP SYMBOLS)
            (PROVES (PROP-AXIOM-PROOF EXP)
                    (F-OR (F-NOT EXP) EXP)
                    GIVEN DEFNS SYMBOLS))
```

At this point in the mechanical proof, we define a series of proof constructors such as PROP-AXIOM-PROOF and prove a series of lemmas such as PROP-AXIOM-PROVES: one for each axiom or proof rule of Z2. Each of the proof constructors such as PROP-AXIOM-PROOF and the definitions of PRF and PROVES are disabled. For example, the proof constructor for the Cut rule is shown below as CUT-PROOF and the corresponding lemma is also displayed as CUT-PROOF-PROVES. The proofs of PROP-AXIOM-PROVES and CUT-PROOF-PROVES only involve the expansion of the relevant definitions and some Boolean simplification, and these are proved quite easily and automatically by the Boyer–Moore theorem prover.

```
Definition.
   (CUT-PROOF A B C PF1 PF2)
      =
   (CONS 2
        (CONS (CONS 4
                   (CONS A (CONS B (CONS C NIL))))
              (CONS (F-OR B C)
                    (CONS (CONS PF1 (CONS PF2 NIL))
                          NIL))))
```

```
Theorem.  CUT-PROOF-PROVES (rewrite):
   (IMPLIES (AND (PROVES PF1
                        (F-OR A B)
                        GIVEN DEFNS SYMBOLS)
                 (AND (PROVES PF2
                             (F-OR (F-NOT A) C)
                             GIVEN DEFNS SYMBOLS)
                      (FORMULA (F-OR B C) SYMBOLS)))
            (PROVES (CUT-PROOF A B C PF1 PF2)
                    (F-OR B C)
                    GIVEN DEFNS SYMBOLS))
```

We now give the first example of the proof of soundness of a derived inference rule. This is the *Commutativity rule for disjunction*. This rule allows us to infer $B \vee A$ from a proof of $A \vee B$. The first step in the proof is to define a proof-constructor COMMUT-PROOF corresponding to this rule.

```
Definition.
   (COMMUT-PROOF A B PF)
      =
   (CUT-PROOF A B A PF
             (PROP-AXIOM-PROOF A))
```

COMMUT-PROOF provides the formal justification for each application of the Commutativity rule and this is given by the following lemma.

```
Theorem.  COMMUT-PROOF-PROVES (rewrite):
   (IMPLIES (PROVES PF
                   (F-OR A B)
                   GIVEN DEFNS SYMBOLS)
            (PROVES (COMMUT-PROOF A B PF)
                    (F-OR B A)
                    GIVEN DEFNS SYMBOLS))
```

The above lemma is also easily proved automatically by expanding the definition of COMMUT-PROOF and rewriting using the the lemmas PROP-AXIOM-PROVES and CUT-PROOF-PROVES. At this point in the mechanical proof, the definition of COMMUT-PROOF is disabled as has been done in the case of PROP-AXIOM-PROOF.

The next important step is the proof of the previously mentioned lemma on disjuncts. The proof of this lemma is nontrivial and our informal exposition follows Shoenfield's *Mathematical Logic* [Sho67, pages 28–29, formula A]. The mechanical proof of this lemma proceeds at roughly the same level of detail as the informal proof in Shoenfield's book. The lemma on disjuncts is an extremely useful derived inference rule. The lemma states that if the formulas $A_1, \ldots, A_m$ are all contained among $B_1, \ldots, B_n$, then we can infer $B_1 \vee \ldots \vee B_n$ from a proof of $A_1 \vee \ldots \vee A_m$. The proof is by induction on m with base cases for $[m = 1]$ and $[m = 2]$. First, we list some of the Lisp functions that are used in the proof but whose definitions are omitted.

- (FORM-LIST FLIST SYMBOLS): Checks if FLIST is a list of formulas using only those function and predicate symbols in the list SYMBOLS.

- (MAKE-DISJUNCT FLIST): Constructs the disjunction of a list of formulas.

- (M1-PROOF EXP FLIST PF): Proof constructor for the $[m = 1]$ case. If PF is a proof of EXP and EXP is a member of FLIST, then M1-PROOF constructs a proof of (MAKE-DISJUNCT FLIST).

- (M2-PROOF EXP1 EXP2 FLIST PF): Proof constructor in the $[m = 2]$ case; it constructs a proof of (MAKE-DISJUNCT FLIST), where EXP1 and EXP2 are members of FLIST and PF is a proof of (F-OR EXP1 EXP2).

- (M-PROOF FLIST1 FLIST2 PF): Constructs a proof of (MAKE-DISJUNCT FLIST2), where the list of formulas FLIST1 is contained in the list of formulas FLIST2 and PF is a proof of (MAKE-DISJUNCT FLIST1).

The lemma M1-PROOF-PROVES1 displayed below expresses the $[m = 1]$ case of the proof. The proof constructor generates the Weakening and Commutativity rule applications required to get a proof of $\ldots \vee A \vee \ldots$ from a proof of A. The parameter CONCL that is used in the lemma is redundant, but it is used because (MAKE-DISJUNCT FLIST) will not directly match against (F-OR A B) when the lemma is used as a rewrite rule.

```
Definition.
  (M1-PROOF EXP FLIST PF)
    =
  (IF (NLISTP FLIST)
      NIL
      (IF (NLISTP (CDR FLIST))
          PF
          (IF (EQUAL EXP (CAR FLIST))
              (RT-EXPAN-PROOF (CAR FLIST)
                              (MAKE-DISJUNCT (CDR FLIST))
                              PF)
              (EXPAN-PROOF (CAR FLIST)
                           (MAKE-DISJUNCT (CDR FLIST))
                           (M1-PROOF EXP (CDR FLIST) PF)))))
```

```
Theorem. M1-PROOF-PROVES1 (rewrite):
   (IMPLIES (AND (FORMULA (MAKE-DISJUNCT FLIST) SYMBOLS)
                 (MEMBER EXP FLIST)
                 (PROVES PF EXP GIVEN DEFNS SYMBOLS)
                 (EQUAL CONCL (MAKE-DISJUNCT FLIST)))
            (PROVES (M1-PROOF EXP FLIST PF)
                    CONCL GIVEN DEFNS SYMBOLS))
```

The [$m = 2$] case of the proof is expressed by the lemma M2-PROOF-PROVES below. EXP1 and EXP2 are the two disjuncts that appear in FLIST. Here, Weakening, Commutativity, and Associativity are used to generate a proof of $\dots \vee A_1 \vee \dots \vee A_2 \vee \dots$ from either a proof of $A_1 \vee A_2$ or a proof of $A_2 \vee A_1$.

```
Definition.
   (M2-PROOF-STEP EXP1 EXP2 FLIST PF)

      =
   (IF (NLISTP FLIST)
       NIL
       (IF (NLISTP (CDR FLIST))
           (IF (EQUAL EXP2 (CAR FLIST)) PF NIL)
           (IF (EQUAL EXP2 (CAR FLIST))
               (RT-ASSOC-PROOF EXP1 EXP2
                               (MAKE-DISJUNCT (CDR FLIST))
                               (RT-EXPAN-PROOF
                                (F-OR EXP1 EXP2)
                                (MAKE-DISJUNCT (CDR FLIST))
                                PF))
               (INSERT-PROOF (CAR FLIST)
                             EXP1
                             (MAKE-DISJUNCT (CDR FLIST))
                             (M2-PROOF-STEP EXP1 EXP2
                                            (CDR FLIST)
                                            PF)))))
```

```
Definition.
   (M2-PROOF EXP1 EXP2 FLIST PF)
      =
   (IF (NLISTP FLIST)
       NIL
       (IF (EQUAL EXP1 EXP2)
           (M1-PROOF EXP1 FLIST
                   (CONTRAC-PROOF EXP1 PF))
          (IF (EQUAL EXP1 (CAR FLIST))
              (M2-PROOF-STEP EXP1 EXP2
                        (CDR FLIST)
                        PF)
             (IF (EQUAL EXP2 (CAR FLIST))
                 (M2-PROOF-STEP EXP2 EXP1
                           (CDR FLIST)
                           (COMMUT-PROOF EXP1 EXP2 PF))
                (EXPAN-PROOF (CAR FLIST)
                      (MAKE-DISJUNCT (CDR FLIST))
                      (M2-PROOF EXP1 EXP2
                            (CDR FLIST)
                            PF)))))))
```

```
Theorem.  M2-PROOF-PROVES (rewrite):
   (IMPLIES (AND (FORMULA (MAKE-DISJUNCT FLIST) SYMBOLS)
                 (MEMBER EXP1 FLIST)
                 (MEMBER EXP2 FLIST)
                 (PROVES PF (F-OR EXP1 EXP2)
                         GIVEN DEFNS SYMBOLS))
            (PROVES (M2-PROOF EXP1 EXP2 FLIST PF)
                    (MAKE-DISJUNCT FLIST)
                    GIVEN DEFNS SYMBOLS))
```

Finally, M-PROOF-PROVES expresses the lemma on disjuncts. If FLIST1 is a list of disjuncts that are contained in FLIST2, and we have a proof of the disjunction of the disjuncts in FLIST1, then M-PROOF constructs a proof of the disjunction of the disjuncts in FLIST2. The proof constructor M-PROOF is quite complex. It recursively invokes itself multiple times and uses Associativity and Contraction to construct the required formal proof. The situation here is that FLIST1 consists of formulas $A_1, \ldots, A_m$ where A is of the form $A_1 \vee \ldots \vee A_m$; FLIST2 consists of formulas $B_1, \ldots, B_n$ where B is of the form $B_1 \vee \ldots \vee B_n$; each A_i is equal to some B_j; and we need to construct a proof of B from A. We already know that M1-PROOF and M2-PROOF take care of the $[m = 1]$ and $[m = 2]$ cases, and we only have to deal with the case when $[m \geq 3]$ by induction. From $\vdash A$, use Associativity to derive $\vdash (A_1 \vee A_2) \vee A_3 \vee \ldots \vee A_m$. Since the latter formula contains $m - 1$ disjuncts, we apply the induction hypothesis to get $\vdash A_1 \vee A_2 \vee B$ which can then be rearranged (using Associativity and Commutativity) as $\vdash (B \vee A_1) \vee A_2$. The $[m = 2]$ case of the proof can be applied to this last formula to get $\vdash (B \vee A_1) \vee B$ which can then be rearranged as $\vdash (B \vee B) \vee A_1$. Similarly applying the $[m = 2]$ case, we can now derive $\vdash (B \vee B) \vee B$ from which $\vdash B$ follows by Contraction.

```
Definition.
   (M-PROOF FLIST1 FLIST2 PF)
      =
   (IF (NLISTP FLIST1)
       NIL
       (IF (NLISTP (CDR FLIST1))
           (M1-PROOF (CAR FLIST1) FLIST2 PF)
           (IF (NLISTP (CDR (CDR FLIST1)))
               (M2-PROOF (CAR FLIST1)
                         (CAR (CDR FLIST1))
                         FLIST2 PF)
               (M3-PROOF
                (CAR FLIST1)
                (CAR (CDR FLIST1))
                FLIST2
                (M-PROOF (CONS (F-OR (CAR FLIST1)
                                     (CAR (CDR FLIST1)))
                              (CDR (CDR FLIST1)))
                         (CONS (F-OR (CAR FLIST1)
                                     (CAR (CDR FLIST1)))
                              FLIST2)
                         (ASSOC-PROOF (CAR FLIST1)
                                      (CAR (CDR FLIST1))
                                      (MAKE-DISJUNCT
                                       (CDR (CDR FLIST1)))
                                      PF))))))
```

```
Theorem.  M-PROOF-PROVES (rewrite):
   (IMPLIES (AND (FORM-LIST FLIST2 SYMBOLS)
                 (LISTP FLIST1)
                 (LISTP FLIST2)
                 (SUBSET FLIST1 FLIST2)
                 (PROVES PF
                         (MAKE-DISJUNCT FLIST1)
                         GIVEN DEFNS SYMBOLS))
            (PROVES (M-PROOF FLIST1 FLIST2 PF)
                    (MAKE-DISJUNCT FLIST2)
                    GIVEN DEFNS SYMBOLS))
```

The remainder of the description of the mechanical proof includes the definition of the tautology checker, the proof of the tautology theorem and the proof of the correctness of the tautology checker.

3.3.2 Defining the Tautology Checker

The tautology checker we define below is an implementation of the one in the informal description of the proof given in Section 3.2. For efficiency reasons, the tautology checker below does not flatten out the entire given formula into a disjunction of atoms or negations of atoms. Instead, it maintains a list of atoms and negations of atoms accumulated so far, and if this list ever contains an atom and its negation, we claim that the given formula is a tautology. A small amount of effort was expended in proving the admissibility of the tautology checker in accordance with the *principle of definition* (page 30) of the Boyer–Moore theorem prover. The steps in the proof of admissibility will be omitted. We describe the preliminary definitions in English and provide skeletal descriptions of the tautology checker before going into the details. First, we define predicates which serve as recognizers for the various classes of formulas. F-ORP recognizes a disjunction. (NOR-TYPE EXP) checks if EXP is of the form $\neg(A \vee B)$. (DBLE-NEG-TYPE EXP) checks if EXP is of the form $\neg\neg A$. If a formula is not of one of the above three forms, then it is either an atom or the negation of an atom.

Next, we examine the definition of the tautology checker TAUTOLOGYP1 in detail. The function TAUTOLOGYP1 takes two arguments, FLIST and AUXLIST. As mentioned earlier, if A is the formula being checked, A must be of the form $A_1 \vee \ldots \vee A_n$, and FLIST is the list $A_1, \ldots, A_n$. The auxiliary argument AUXLIST accumulates the atoms and negations of atoms encountered during the recursion of the tautology checker. AUXLIST will have to be bound to NIL initially, when invoking TAUTOLOGYP1. (ARG EXP) returns B if EXP is of the form $\neg B$. (ARG1 EXP) returns B when given an expression EXP of the form $\neg B$ or of the form $B \vee C$, and in the latter case, (ARG2 EXP) returns C. The definition of TAUTOLOGYP1 is displayed below. If FLIST is empty, we return F. Otherwise we check if the first element of FLIST is of one of the types: F-ORP, NOR-TYPE, DBLE-NEG-TYPE, or an atom or negation of an atom, and branch accordingly:

- When (F-ORP (CAR FLIST)) is T, the first element of FLIST is a disjunction of the form $B \vee C$, we then add B and C to the rest of the FLIST and recursively invoke TAUTOLOGYP1 with the AUXLIST unchanged.

- When (NOR-TYPE (CAR FLIST)), the first element of FLIST is of the form $(\neg B \vee C)$, then there are two recursive calls to TAUTOLOGYP1: one with $\neg B$ added to the rest of FLIST, and another with $\neg C$ added to the rest of FLIST.

- When (DBLE-NEG-TYPE (CAR FLIST)), the first element is of the form $\neg B$, then B is added to the rest of FLIST and a recursive call is made.

- The only remaining possibility is that the first element of FLIST is either an atom or the negation of an atom. In this case, TAUTOLOGYP1 first checks if the negation of the (CAR FLIST) appears in AUXLIST and returns T if that is the case. Otherwise, the first element of FLIST is added to the AUXLIST and TAUTOLOGYP1 is recursively invoked on the rest of the FLIST. The Lisp expression (NEG-LIST EXP FLIST) checks if either EXP is the negation of some member of FLIST or if some member of FLIST is the negation of EXP.

```
Definition.
  (TAUTOLOGYP1 FLIST AUXLIST)
    =
  (IF
   (LISTP FLIST)
   (IF
    (F-ORP (CAR FLIST))
    (TAUTOLOGYP1 (CONS (ARG1 (CAR FLIST))
                       (CONS (ARG2 (CAR FLIST)) (CDR FLIST)))
                 AUXLIST)
    (IF (NOR-TYPE (CAR FLIST))
        (AND (TAUTOLOGYP1 (CONS (F-NOT
                                  (ARG1 (ARG (CAR FLIST))))
                                (CDR FLIST))
                          AUXLIST)
             (TAUTOLOGYP1 (CONS (F-NOT
                                  (ARG2 (ARG (CAR FLIST))))
                                (CDR FLIST))
                          AUXLIST))
        (IF (DBLE-NEG-TYPE (CAR FLIST))
            (TAUTOLOGYP1 (CONS (ARG (ARG (CAR FLIST)))
                               (CDR FLIST))
                         AUXLIST)
            (OR (NEG-LIST (CAR FLIST) AUXLIST)
                (TAUTOLOGYP1 (CDR FLIST)
                             (CONS (CAR FLIST) AUXLIST))))))
   F)
```

This concludes the description of the tautology checker for Shoenfield's first-order logic. Two examples of invocations of TAUTOLOGYP1 are presented below. The term

```
(TAUTOLOGYP1 (LIST (F-IMPLIES (F-EQUAL (V 0) (V 1))
                   (F-IMPLIES (F-EQUAL (V 3)(V 2))
                              (F-EQUAL (V 0)(V 1)))))

             NIL)
```

evaluates to T. On the other hand, the term

```
(TAUTOLOGYP1 (LIST (F-IMPLIES (F-EQUAL (V 0)(V 1))
                   (F-EQUAL (V 1)(V 0))))

             NIL)
```

evaluates to F. Later, in Section 3.3.4, TAUTOLOGYP1 will be proved correct with respect to the truth-table definition of a tautology.

3.3.3 The Proof of the Tautology Theorem

We sketch below the mechanical proof of the statement that all tautologies have formal proofs within first-order logic. The major task in the proof is to define the proof-constructor function which constructs formal proofs for those formulas on which the tautology checker returns T. More accurately, the proof constructor TAUT-PROOF1 constructs a proof of (MAKE-DISJUNCT

(APPEND FLIST AUXLIST)), where APPEND concatenates two lists, and MAKE-DISJUNCT returns the disjunction of a list of formulas. So when AUXLIST is NIL, the proof constructor constructs a proof of (MAKE-DISJUNCT FLIST). The case structure and recursion scheme employed by the proof constructor TAUT-PROOF1 are identical to those of TAUTOLOGYP1. The control skeleton of TAUT-PROOF1 is displayed below.

```
Definition.
  (TAUT-PROOF1 FLIST AUXLIST)
    =
  (IF (LISTP FLIST)
      (IF (F-ORP (CAR FLIST))
          {proof constructor for F-ORP case}
          (IF (NOR-TYPE (CAR FLIST))
              {proof constructor for NOR-TYPE case}
              (IF (DBLE-NEG-TYPE (CAR FLIST))
                  {proof constructor for DBLE-NEG-TYPE case}
                  {proof constructor for atom/negation of
                   atom case})))
      NIL)
```

The body of TAUT-PROOF1 makes calls to several other proof constructors and we omit several lemmas which state that these functions construct the appropriate proofs. The formal proofs constructed by these proof constructors will be described in the text. (APPEND (CDR FLIST) AUXLIST) when nonempty is labelled C. (CAR FLIST) can either be an atom, a negation of an atom, or of one of the forms: $A \vee B$, $\neg(A \vee B)$, or $\neg(\neg A)$.

In the F-ORP case, FLIST is of the form $\{(A \vee B), \ldots\}$, and F-ORP-PROOF constructs the proof of the disjunction of the formulas in FLIST and AUXLIST using the proof generated by the recursive call to TAUT-PROOF1. If (APPEND (CDR FLIST) AUXLIST) is empty, the recursive call returns a required proof of $(A \vee B)$. Otherwise, the recursive call to TAUT-PROOF1 returns a proof of $(A \vee (B \vee C))$ and one application of the Associativity rule leads to the required proof.

```
(F-ORP-PROOF (ARG1 (CAR FLIST))
             (ARG2 (CAR FLIST))
             (APPEND (CDR FLIST) AUXLIST)
             (TAUT-PROOF1 (CONS (ARG1 (CAR FLIST))
                                (CONS (ARG2 (CAR FLIST))
                                      (CDR FLIST)))
                          AUXLIST))
```

NOR-PROOF constructs the proof in the NOR-TYPE case but this time there are two recursive calls to TAUT-PROOF1 as is also the case in TAUTOLOGYP1. If (APPEND (CDR FLIST) AUXLIST) is empty, the recursive calls return proofs of $\neg A$ and $\neg B$, respectively. We use a cancellation rule which permits the derivation of E from $\neg D$ and $D \vee E$. First we derive $(A \vee (B \vee \neg(A \vee B)))$ from the propositional axiom $\neg(A \vee B) \vee (A \vee B)$ by Lemma 3.2.2. Then we cancel A and B in turn to get the desired conclusion. In the case when (APPEND (CDR FLIST) AUXLIST) is nonempty, the recursive calls return proofs of $(\neg A \vee C)$ and $(\neg B \vee C)$, respectively. As before we get $(A \vee (B \vee \neg(A \vee B)))$ and apply the Cut rule with $(\neg A \vee C)$, and Associativity to get $(B \vee (\neg(A \vee B) \vee C))$. One more application of the Cut rule yields $(\neg(A \vee B) \vee C \vee C)$ from which we get the desired result $(\neg(A \vee B) \vee C)$ by Lemma 3.2.2.

```
(NOR-PROOF (ARG1 (ARG (CAR FLIST)))
           (ARG2 (ARG (CAR FLIST)))
           (APPEND (CDR FLIST) AUXLIST)
           (TAUT-PROOF1
             (CONS (F-NOT (ARG1 (ARG (CAR FLIST))))
                   (CDR FLIST))
             AUXLIST)
           (TAUT-PROOF1
             (CONS (F-NOT (ARG2 (ARG (CAR FLIST))))
                   (CDR FLIST))
             AUXLIST))
```

In the DBLE-NEG-TYPE case, DBLE-NEG-PROOF is used to construct the required proof from the proof generated by the recursive call to TAUT-PROOF1. When (APPEND (CDR FLIST) AUXLIST) is empty, the recursive call generates a proof of A from which we can derive $(A \lor (\neg(\neg A)))$. An application of the Cut rule with $((\neg A) \lor (\neg(\neg A)))$ (obtained by Commutativity from a propositional axiom) then yields $((\neg(\neg A)) \lor (\neg(\neg A)))$ which by the Contraction rule yields the desired result. In the case when (APPEND (CDR FLIST) AUXLIST) is nonempty, we instead apply the Cut rule with $\neg A \lor \neg\neg A$ and $(A \lor C)$ (got from the recursive call), to derive the result $\neg\neg A \lor C$.

```
(DBLE-NEG-PROOF (ARG (ARG (CAR FLIST)))
                (APPEND (CDR FLIST) AUXLIST)
                (TAUT-PROOF1
                  (CONS (ARG (ARG (CAR FLIST)))
                        (CDR FLIST))
                  AUXLIST))
```

In the remaining case, two possibilities arise depending on whether (NEG-LIST (CAR FLIST) AUXLIST) is T or not. If it is T, then PROP-ATOM-PROOF1 constructs the appropriate proof by applying the Lemma 3.2.2 to the propositional axiom $\neg A \lor A$, where A is (CAR FLIST). If it is not T, then we employ a recursion scheme that is similar to that of TAUTOLOGYP1, so that PROP-ATOM-PROOF2 constructs the required final proof using Lemma 3.2.2.

```
(IF (NEG-LIST (CAR FLIST) AUXLIST)
    (PROP-ATOM-PROOF1 FLIST AUXLIST)
    (PROP-ATOM-PROOF2 FLIST AUXLIST
                      (TAUT-PROOF1
                        (CDR FLIST)
                        (CONS (CAR FLIST)
                              AUXLIST))))
```

We now state the theorem which asserts that TAUT-PROOF1 constructs a correct proof of (MAKE-DISJUNCT (APPEND FLIST AUXLIST)) if (TAUTOLOGYP1 FLIST AUXLIST) is T and both FLIST and AUXLIST are lists of formulas.

```
Theorem.  TAUT-THM1 (rewrite):
   (IMPLIES (AND (FORM-LIST FLIST SYMBOLS)
                 (FORM-LIST AUXLIST SYMBOLS)
                 (TAUTOLOGYP1 FLIST AUXLIST))
            (PROVES (TAUT-PROOF1 FLIST AUXLIST)
                    (MAKE-DISJUNCT (APPEND FLIST AUXLIST))
                    GIVEN DEFNS SYMBOLS))
```

The theorem TAUT-THM1 captures the statement of the tautology theorem when AUXLIST is instantiated with NIL.

3.3.4   The Correctness of the Tautology Checker

The final part of the proof consists in showing that the tautology checker TAUTOLOGYP1 corresponds to the truth-table definition of a tautology. Boyer and Moore [BM79] have carried out a similar proof of the correctness of a tautology checker for IF-expressions. To prove the correctness of the above tautology checker, we need to:

1. Define a function which evaluates the truth value of a formula with respect to a given truth assignment.

2. Using the above function, show that if TAUTOLOGYP1 asserts a given formula to be a tautology, then the truth value of that formula under any truth assignment is always **true**.

3. Prove that if TAUTOLOGYP1 claims that the given formula is not a tautology, a falsifying truth assignment exists, that is, an assignment under which the truth value of the given formula is **false**.

The function EVAL below evaluates the truth value of the formula EXP with respect to the truth assignment ALIST and returns T or F accordingly. EVAL works as follows. If EXP is of the form $\neg A$, then EVAL returns the negation of the truth value of A with respect to ALIST. If EXP is of the form $A \lor B$, then EVAL returns T if at least one of A or B evaluates to T with respect to ALIST, otherwise EVAL returns F. In the remaining case EXP is an atom, and EVAL returns T if EXP is a member of ALIST and F otherwise.

```
Definition.
   (EVAL EXP ALIST)
     =
   (IF (F-NOTP EXP)
       (NOT (EVAL (ARG EXP) ALIST))
       (IF (F-ORP EXP)
           (OR (EVAL (ARG1 EXP) ALIST)
               (EVAL (ARG2 EXP) ALIST))
           (MEMBER EXP ALIST)))
```

Having defined EVAL, we can state and prove the other two statements in the proof of correctness of TAUTOLOGYP1. Since free variables like ALIST are implicitly universally quantified, the theorem TAUT-EVAL states that if (TAUTOLOGYP1 FLIST AUXLIST) is T, then EVAL on (MAKE-DISJUNCT (APPEND FLIST AUXLIST)) returns T on any ALIST.

```
Theorem.  TAUT-EVAL (rewrite):
   (IMPLIES (TAUTOLOGYP1 FLIST AUXLIST)
            (EVAL (MAKE-DISJUNCT (APPEND FLIST AUXLIST))
                  ALIST))
```

(FALSIFY-TAUT FLIST AUXLIST) explicitly constructs the truth assignment which falsifies (MAKE-DISJUNCT (APPEND FLIST AUXLIST)) when (TAUTOLOGYP1 FLIST AUXLIST) is F. Its actual definition is not needed to convince ourselves that the statement of the theorem ensures that non-tautologies are falsifiable. We state this fact as NOT-TAUT-FALSE below. We ensure that AUXLIST is a list of propositional atoms which does not contain both an atom and its negation. This is checked by PROP-ATOMP-LIST. Since we are only interested in the instance when the AUXLIST is NIL, the above restriction turns out not to matter.

```
Theorem.  NOT-TAUT-FALSE (rewrite):
   (IMPLIES (AND (PROP-ATOMP-LIST AUXLIST)
                 (NOT (TAUTOLOGYP1 FLIST AUXLIST)))
            (NOT (EVAL (MAKE-DISJUNCT (APPEND FLIST AUXLIST))
                       (FALSIFY-TAUT FLIST AUXLIST))))
```

Finally, we replace AUXLIST by NIL and derive more readable versions of the above theorems.

```
Definition.
   (TAUTOLOGYP FLIST)
      =
   (TAUTOLOGYP1 FLIST NIL)
```

```
Definition.
   (TAUT-PROOF FLIST)
      =
   (TAUT-PROOF1 FLIST NIL)
```

```
Theorem.  TAUTOLOGY-THEOREM (rewrite):
   (IMPLIES (AND (FORM-LIST FLIST SYMBOLS)
                 (TAUTOLOGYP FLIST)
                 (EQUAL CONCL (MAKE-DISJUNCT FLIST)))
            (PROVES (TAUT-PROOF FLIST)
                    CONCL GIVEN DEFNS SYMBOLS))
```

```
Theorem.  TAUTOLOGIES-ARE-TRUE (rewrite):
   (IMPLIES (AND (LISTP FLIST) (TAUTOLOGYP FLIST))
            (EVAL (MAKE-DISJUNCT FLIST) ALIST))
```

```
Theorem.  TRUTHS-ARE-TAUTOLOGIES (rewrite):
   (IMPLIES (NOT (TAUTOLOGYP FLIST))
            (NOT (EVAL (MAKE-DISJUNCT FLIST)
                       (FALSIFY-TAUT FLIST NIL))))
```

3.3.5   Using the Tautology Theorem

The main motivation for proving TAUTOLOGY-THEOREM was that it could then be applied to simplify some of the formal deduction steps in the metatheorems to follow. As it turned out, it was not directly usable since all our applications involve the use of metavariables, that is, variables in the Boyer–Moore logic, and the tautology checker can only handle formal expressions. For instance, if we want to show that (F-OR (F-NOT A) A) is a tautology for any formula A, we cannot directly apply the tautology checker without instantiating A. The tautology theorem was rendered useful by the contrapositive version of TRUTHS-ARE-TAUTOLOGIES displayed below as EVAL-TAUTOLOGYP. Since EVAL translates formal disjunctions and negations into disjunctions and negations in the Boyer–Moore logic, we can use it to translate tautologies containing metavariables into formulas that are tautologously true in the Boyer–Moore logic. Now, in order to apply TAUTOLOGY-THEOREM, the theorem prover attempts to establish (TAUTOLOGYP FLIST), which by EVAL-TAUTOLOGYP reduces to showing that (EVAL (MAKE-DISJUNCT FLIST) (FALSIFY-TAUT FLIST NIL)) holds.

```
Theorem.  EVAL-TAUTOLOGYP (rewrite):
   (IMPLIES (EVAL (MAKE-DISJUNCT FLIST)
                  (FALSIFY-TAUT FLIST NIL))
            (TAUTOLOGYP FLIST))
```

A common way in which tautologies arise in proofs is when there are n proven formulas $\{A_1, ..., A_n\}$, and we wish to prove B, where B is a *tautological consequence* of $\{A_1, ..., A_n\}$, that is, $(A_1 \supset (A_2 \supset ... (A_n \supset B)))$ is a tautology. The use of this form of the tautology theorem is made possible by the following derived inference rule which is a consequence of EVAL-TAUTOLOGYP and TAUTOLOGY-THEOREM. It asserts that if EXP is a formula which is a tautological consequence of the formulas in FLIST, then EXP is provable if each of the formulas in FLIST is provable. The operation LIS-NOT negates each formula in a list of formulas.

```
Theorem.  EVAL-TAUTCONSEQ-PROOF-PROVES (rewrite):
   (IMPLIES (AND (FORMULA EXP SYMBOLS)
                 (EVAL (MAKE-DISJUNCT (APPEND (LIS-NOT FLIST)
                                             (CONS EXP
                                                   NIL)))
                       (FALSIFY-TAUT (APPEND (LIS-NOT FLIST)
                                            (CONS EXP NIL))
                                     NIL))
                 (PROVES-LIST PFLIST FLIST GIVEN
                              DEFNS SYMBOLS))
            (PROVES (TAUTCONSEQ-PROOF FLIST EXP PFLIST)
                    EXP GIVEN DEFNS SYMBOLS))
```

This is perhaps the most frequently employed lemma in the metatheorems that follow. Several properties of the logical connectives $\neg$, $\vee$, $\wedge$, $\supset$, $\Longleftrightarrow$, and their inter-relationships are proved using this lemma. A few examples of these are:

1. The transitivity of $\supset$: derive $(A \supset C)$ from $(A \supset B)$ and $(B \supset C)$.

2. The rule of contraposition: derive $(\neg B \supset \neg A)$ from $(A \supset B)$.

3. $\wedge$-introduction: derive $(A \wedge B)$ from A and B.

This concludes the discussion of the tautology theorem, its mechanical proof, and the various consequences of the tautology theorem.

3.4 Other Derived Inference Rules

In the course of the proof of the incompleteness theorem, several derived inference rules were proved as metatheorems and used to demonstrate the existence of formal proofs. We examine a few of these below. They include the instantiation rule, the equivalence theorem, and the equality theorem.

3.4.1 The Instantiation Rule

The instantiation rule states that any *instance* of a theorem is a theorem. This is usually taken as a primitive inference rule, but is proved as a derived rule in Shoenfield's logic. The following simple rules are first proved.

1. $\forall$-Introduction: derive $(A \supset ((\forall x.\ B)))$ from $(A \supset B)$, provided x does not occur free in A.

2. Generalization Rule: derive $(\forall x.\ A)$ from A.

3. Substitution Rule: derive $[t/x]A$ from A, if **true** is free for x in A.

B is an *instance* of A if B is got from A by simultaneously substituting the t_i from some list of terms $t_1, ..., t_n$ for the corresponding x_i from some list of distinct variables $x_1, ..., x_n$. Each t_i must be free for the corresponding x_i in A. Such an instance of A will be denoted as $[t_1/x_1, \ldots, t_n/x_n]A$. The manner in which simultaneous substitution is carried out is important. The explanation above seems to indicate that it is performed by recursion on the structure of A and replaces each free occurrence of an x_i by the corresponding t_i. However, since the instantiation rule must be derived from the single substitution rule above, simultaneous substitution must also be defined in terms of several single substitutions. This definition has the problem that if we first substitute t_1 for x_1, some free occurrence of an x_i in t_1 could be replaced in a later substitution. To prevent this, we first need to rename all the x_i in A by some entirely new variables y_i. Then we can safely substitute t_i for y_i in order, since no y_j occurs in any t_k. This involves n substitutions to replace the x_i with the y_i, and another n substitutions to replace the y_i with t_i. This becomes inefficient for large A. We therefore prove that the two methods of performing simultaneous substitution return the same answer. The mechanical proof of this equivalence is surprisingly succinct.

3.4.2 The Equivalence Theorem

Another commonly used rule of inference in informal proofs is the equivalence theorem which permits the replacement of a subformula by an equivalent one. Suppose we have a list of proven equivalences $(A_1 \iff B_1), ..., (A_n \iff B_n)$, and A and B are the same except that some occurrences of an A_i in A might be replaced by the corresponding B_i in B. Then we can derive a proof of $(A \iff B)$. To prove this derived inference rule we define an "equivalence checker" by simultaneous recursion on the structures of A and B that checks if A and B are the same except that some A_i in A are replaced by the corresponding B_i in B. The proof is by an induction similar to the recursion above, and uses the fact that equivalence is preserved under the operations of negation, disjunction, and existential quantification.

3.4.3   The Equality Theorem

The equality theorem is similar to the equivalence theorem and captures the intuitive operation of "replacing equals for equals." This theorem asserts that if $(a_1 = b_1), ..., (a_n = b_n)$ are proven equalities, and A and B are the same except that some occurrences of an a_i in A might be replaced by the corresponding b_i in B, then we can derive a proof of $(A \iff B)$. The equality theorem is also proved by first defining an "equality checker" and performing an induction corresponding to the recursion displayed by the equality checker. We use as a lemma the corresponding theorem for terms where we show that $(a = b)$ is provable, when a and b are identical except that some occurrences of a_i in a are replaced by the corresponding b_i in b.

3.5   Summary

The present chapter dealt with the proofs of the soundness of various derived inference rules, mainly the proof of the tautology theorem. The role of the mechanical proof was to certify that such derived inference rules are sound extensions to the proof checker described in Chapter 2. Though these proofs are quite complicated, they were checked at a fairly high-level using the Boyer–Moore theorem prover. The theorem prover was given approximately the same information as one would glean from a careful reading of an informal description of the proof. Many complicated lemmas, especially several ones regarding the substitution operation, were proved by the theorem prover without any assistance.

In our description of the mechanical proof of Gödel's incompleteness theorem, we have so far outlined

1. The definition of a proof checker for proofs written in the formal logic Z2.

2. The statement of Gödel's incompleteness theorem that Z2 is either incomplete or inconsistent. The proof checker forms a crucial part of this statement.

3. The extension of this proof checker with significant derived inference rules. These make the task of constructing and checking formal proofs manageable. This ability to metamathematically extend proof checkers makes it possible to check formal proofs at a much higher level and with greater efficiency.

In order to complete the outline, we need to construct an undecidable sentence in Z2. This sentence is constructed by representing the metatheory of Z2 in Z2 itself and constructing a sentence that asserts its own unprovability.

Chapter 4

The Representability of the Metatheory

> *We shall give only an outline of the proof of this theorem because the proof does not present any difficulty in principle and is rather long.*
>
> Kurt Gödel [Göd67b]

The climactic step in Gödel's proof of the incompleteness theorem is the construction of a sentence asserting its own unprovability. Since provability is a metatheoretic notion, the construction of such a sentence within Z2 requires that some part of the metatheory of Z2 be represented within Z2 itself. The main result of this chapter is an outline of the proof of the representability of the metatheory of Z2 (as described in Chapter 2) within Z2 . The formalization of the representability theorem is perhaps the single most substantial and difficult part of our mechanical verification. The machinery of derived inference rules developed in Chapter 3 is heavily exploited in the representability proofs. The representability theorem is used in the construction of the undecidable sentence in Chapter 5.

The contents of this and the succeeding chapter are quite complicated since they define and use a seemingly unmanageable variety of encodings. One encoding represents the syntax of Z2 expressions using only numbers and the pairing operation of Lisp, and another encoding represents these latter Lisp data structures as sets in Z2. The representability theorem then shows the correspondence (under the encoding) between the metatheoretic Lisp computations and certain kinds of proofs in Z2.

4.1 A Brief Overview

We saw in Chapter 2 how the metatheory of Z2 could be formalized in terms of Lisp functions such as COLLECT-FREE, SUBST, PRF, and PROVES. This chapter shows how the evaluation of these Lisp functions can be simulated within Z2. While the basic idea is quite simple, the task of fleshing out the technical details can be quite demanding. The Gödel-encoding of the syntax of Z2 as Z2 sets is straightforward: numbers are mapped to numerals, and CONSes are mapped to ordered pairs. Let ⌈X⌉ represent the Gödel-encoding of a Lisp data structure X. Suppose that g is some unary Lisp function. The representability theorem asserts that to any such Lisp function, there is a Z2 formula A_g with exactly two free variables, say x and y, such that:

1. If $(g \ X) = Y$, then $\vdash [\lceil X \rceil/x, \lceil Y \rceil/y]A_g$, and

2. If $(g \ X) \neq Y$, then $\vdash [\lceil X \rceil/x, \lceil Y \rceil/y]\neg A_g$.

The above statement can easily be generalized to Lisp functions of multiple arguments to get the general form of the representability theorem shown below.

$$\text{If } (g \ X_1 \ \ldots \ X_n) = Y,$$
$$\text{then} \quad \vdash [\lceil X_1 \rceil/x_1, \ldots, \lceil X_n \rceil/x_n, \lceil Y \rceil/y]A_g. \tag{4.1}$$
$$\text{If } (g \ X_1 \ \ldots \ X_n) \neq Y,$$
$$\text{then} \quad \vdash [\lceil X_1 \rceil/x_1, \ldots, \lceil X_n \rceil/x_n, \lceil Y \rceil/y]\neg A_g. \tag{4.2}$$

It is easy to see that simple functions like CONS, CAR, ADD1, SUB1, and so on, are representable by the corresponding operations on numerals and ordered pairs. In fact, the Z2 analogues of these operations are defined in Section 4.4.1 and their definitions are shown to be admissible in Z2. It is harder to see how recursive Lisp functions like APPEND, COLLECT-FREE, and especially the proof-checking functions PRF and PROVES, are representable. The trick with representing recursive functions is to define the notion of a *trace* corresponding to the computation of the function. A *trace* for the Lisp function g on the argument X would contain $\langle \lceil (g \ Z) \rceil, \lceil Z \rceil \rangle$, for any Z such that $(g \ Z)$ is evaluated during the computation of $(g \ X)$. For example, the trace for (LENGTH '(1 . 0)) would be the set:

$$\{\langle \lceil 1 \rceil, \lceil (\text{CONS } 1 \ 0) \rceil \rangle, \quad \langle \lceil 0 \rceil, \lceil 0 \rceil \rangle\}.$$

The representing formula A_g, roughly speaking, asserts the existence of a unique such trace.

The representability of an individual Lisp function g is proved by constructing a trace corresponding to a given argument X and arguing that every other acceptable trace assigns the same value to the function at the argument X. Lisp functions of several arguments can be similarly shown to be representable. Informal presentations of the representability proof show the details of the representability of one or two metatheoretic functions and argue that the other functions can be similarly represented. Such an argument is acceptable for an informal exposition where the reader can be urged to grasp the general pattern, but is unacceptable for a formalized proof. In Gödel's original paper [Göd67b], for example, the metatheory is presented as a series of primitive recursive definitions, and all primitive recursive functions are shown to be representable in the object theory. In order to observe that all the metatheoretic functions are primitive recursive, one must be in the meta-metatheory.

The proof here has been formalized in the metatheory itself. The representability argument is simplified by demonstrating the representability of a Lisp interpreter, a Lisp function that evaluates any given Lisp expression. The representability theorem is just the representability of this Lisp interpreter. It is shown that each of the Lisp functions used in the metatheoretic description of Z2 can be evaluated using the Lisp interpreter. The computability of the metatheory by the interpreter and the representability of the interpreter taken together, yield the representability of the metatheory. This subtle and seemingly innocuous use of a Lisp interpreter is perhaps the single most important "design decision" in the mechanical verification.

The representability theorem is presented below in several steps. The first step in Section 4.2 is to define a Lisp function that is itself an evaluator for Lisp functions. Note that this is possible because the syntax of Lisp can be easily represented using the data structures of Lisp. The second step in Sections 4.3 and 4.4 is to establish the representability of the Lisp evaluator in Z2. The representability proof makes heavy use of the derived inference rules established in Chapter 3. The final step in Section 4.5 is to demonstrate that the metatheoretic Lisp functions of Z2 can be computed using the Lisp evaluator. These three steps together yield the embedding of the metatheory of Z2 within the theory of Z2 thus making it possible to construct a self-referential undecidable sentence.

4.2   The Lisp Interpreter

The task here is to define a Lisp interpreter or evaluator in Lisp itself. Such an interpreter was first defined by McCarthy as a theoretical curiosity but soon became the basis for an early implementation of Lisp [MAE+65, McC78]. To define such an interpreter within Lisp, we represent Lisp expressions as Lisp data structures. These representations of Lisp expressions are then evaluated with respect to a list of bindings of values to variables and function definitions to function symbols. Some care is needed to define this interpreter so that it is admissible under the principle of definition employed by the Boyer–Moore theorem prover. Note that the expressions evaluated by the interpreter includes those whose evaluation need not terminate, such as the Lisp function BAD, where (BAD X) is defined as (NOT (BAD X)).

The interpreter defined here is shown in Section 4.4 to be representable in Z2. The metatheoretic functions in Chapter 2 are shown in Section 4.5 to be computable using this interpreter and are therefore representable in Z2. The representability of these metatheoretic functions is used to construct the required undecidable sentence in Chapter 5.

We will define a Lisp function EV as the Lisp interpreter. This function evaluates a given Lisp object representing a Lisp expression. The given Lisp expression to be evaluated can contain Lisp variables. These variables get their values from an argument to EV that attaches values to variables. Yet another argument to EV attaches definitions to the function symbols used in the given Lisp expression. Both variables and function symbols are represented by numbers. The application of a function FN to argument expressions ARG1 to ARGN is represented by the list (FN ARG1 ... ARGN). To avoid any confusion with the function and variable symbols in Z2, the functions and variables in the expressions evaluated by EV will be referred to as *e-functions* and *e-variables* ('e' for EV), respectively. The class of Lisp objects representing Lisp expressions will be referred to as *e-expressions*.

Note that in order to simplify the proof of the representability of EV in Z2, the syntax of e-expressions has been defined solely in terms of the natural numbers and the CONS constructor of Lisp. This means that any non-numeric or non-CONS e-expression is treated as the e-expression 0. We call any Lisp object that is constructed solely from numbers and lists an *n-expression*. The n-expressions thus consist of the Lisp numerals, and ordered pairs of the form (CONS a b) where a and b are n-expressions. Not only is the syntax of e-expressions represented by Lisp n-expressions but so are the values obtained from evaluating these e-expressions. Thus, if 7 is the e-function representing CONS and 10 is the e-function rep-

resenting QUOTE, then the e-expression (7 (10 1) (10 2)) represents the Lisp expression
(CONS '1 '2), and evaluates to the n-expression (1 . 2).[1]

The Lisp equality predicate restricted to n-expressions is given by NCODE-EQUAL below so
that all non-n-expressions are equivalent to each other and to 0.

```
Definition.
  (NCODE-EQUAL X Y)
    =
  (IF (LISTP X)
      (IF (LISTP Y)
          (AND (NCODE-EQUAL (CAR X) (CAR Y))
               (NCODE-EQUAL (CDR X) (CDR Y)))
          F)
      (IF (LISTP Y)
          F
          (IF (ZEROP X)
              (ZEROP Y)
              (IF (ZEROP Y)
                  F
                  (NCODE-EQUAL (SUB1 X) (SUB1 Y)))))))
```

The evaluator EV can be invoked on e-expressions whose evaluations do not terminate, but
there is a bound on the evaluation of EV determined by an extra argument to the function. If
this bound is not large enough to compute the value of the given e-expression, then a value
(BTM) is returned by EV to represent an undefined or undetermined value. This special value
is introduced by the shell definition shown below which introduces a shell containing (BTM)
and nothing else.

```
Shell Definition.
  Add the shell BTM of zero arguments with
  recognizer BTMP,
  accessors
  and default values
```

The definition of EV employs the function GET to extract the binding corresponding to an
e-variable or an e-function symbol. (GET X Y) is defined to return the X+1st element of the
list Y. Note that if the list Y is empty, GET returns the value 0 since (CAR X) is 0 when X is
an empty list.

```
Definition.
  (GET X Y)
    =
  (IF (ZEROP X)
      (CAR Y)
      (GET (SUB1 X) (CDR Y)))
```

---

[1]The use of the same representation (n-expressions) for e-expressions as well as their values is of course
highly confusing, but this confusion is equally present in Lisp itself. As the example shows, the QUOTE operation
is used to express Lisp values (such as the number 2) as Lisp expressions as in '2. In Lisp, numbers are self-
evaluating and do not need to be quoted, but in an e-expression, numbers are treated as variables so that an
e-expression representing a number, say 2, has to be in the quoted form (10 2).

The e-function symbols represent either the primitive operations or those that are defined in an environment given as an argument to EV. The primitive e-function symbols are represented by the Lisp numerals from 0 to 10. The numeral 0 represents the IF conditional. The numeral 10 represents Lisp's QUOTE operation. As explained below, both these operations are evaluated in a special manner by EV. The remaining numerals from 1 through 9 respectively represent the primitive Lisp functions EQUAL, NUMBERP, ZEROP, ADD1, SUB1, LISTP, CONS, CAR, and CDR. The evaluation of these primitive e-functions on a given list of argument values ARGS is described by the list of definitions below leading up to APPLY-SUBR. The BOOL-FIX function converts F to 0 and coerces the remaining values to 1 since numbers are used instead of the booleans in the e-expressions.

```
Definition.
  (BOOL-FIX X)
     =
  (IF X 1 0)
```

APPLY-SUBR1 evaluates the e-functions representing the list operations LISTP, CONS, CAR, and CDR.

```
Definition.
  (APPLY-SUBR1 FN ARGS)
     =
  (IF (EQUAL FN 6)
      (BOOL-FIX (LISTP (CAR ARGS)))
      (IF (EQUAL FN 7)
          (CONS (CAR ARGS) (CAR (CDR ARGS)))
          (IF (EQUAL FN 8)
              (CAR (CAR ARGS))
              (CDR (CAR ARGS)))))
```

APPLY-SUBR2 evaluates the e-functions representing the numeric operations ZEROP, ADD1, and SUB1.

```
Definition.
  (APPLY-SUBR2 FN ARGS)
     =
  (IF (EQUAL FN 3)
      (BOOL-FIX (ZEROP (CAR ARGS)))
      (IF (EQUAL FN 4)
          (ADD1 (CAR ARGS))
          (IF (EQUAL FN 5)
              (SUB1 (CAR ARGS))
              (APPLY-SUBR1 FN ARGS))))
```

APPLY-SUBR evaluates the e-functions for EQUAL and NUMBERP. The e-function for NUMBERP is defined here to recognize all those values on which LISTP is F.

```
Definition.
  (APPLY-SUBR FN ARGS)
    =
  (IF (EQUAL FN 1)
      (BOOL-FIX (NCODE-EQUAL (CAR ARGS)
                             (CAR (CDR ARGS))))
      (IF (EQUAL FN 2)
          (BOOL-FIX (NOT (LISTP (CAR ARGS))))
          (APPLY-SUBR2 FN ARGS)))
```

Thus, e-expressions are either:

- e-variables (represented as natural numbers), or

- lists of the form (CONS FN ARGS), where FN is the e-function symbol (also represented by a natural number) and ARGS is a list of argument e-expressions.

The interpreter EV takes five arguments:

1. FLG: This is either 0 or 1 according to whether a single e-expression or a list of e-expressions is being evaluated.

2. EXP: This is the e-expression to be evaluated when FLG is 0, or is a list of e-expressions when FLG is 1.

3. VA: This is the *variable environment* that assigns values to e-variables. Since e-variables are represented by numbers, VA is just a list of values so that the value of the e-variable i is the $i + 1$st element of VA.[2]

4. FA: This is the list of e-function definitions and, as with VA, the number used to represent the e-function symbol determines the position of its definition in FA.

5. N: This argument is needed to place a bound on the size of the computation in order to ensure that the definition of EV is admissible in the Boyer–Moore logic. In order to properly evaluate a given e-expression, one must guess a high enough value for this bound. Otherwise EV returns (BTM) to indicate that the evaluation has not terminated within the given bound.

The definition of EV can now be presented. When FLG is 1, EXP is a list of e-expressions and each of these expressions is evaluated. If any of the e-expressions in the given list evaluates to (BTM), then EV returns the value (BTM). Otherwise, EV returns the list of values obtained by evaluating each of the e-expressions in the given list in turn.

When FLG is not 1 and by convention set to 0, then EXP is a single e-expression and the following cases arise:

- EXP is not a list. Then it represents an e-variable; EV returns (GET EXP VA) as the value.

---

[2] The use of numbers to represent variables and functions by their position in the binding environment is inspired by the de Bruijn representation for lambda calculus which is extensively discussed in Chapter 6. The significance of these simplifying representations cannot be emphasized enough: they greatly reduce the amount of labor and can make the difference between an infeasible and feasible mechanical verification.

- **EXP** is a list and therefore of the form (CONS FN ARGS), where FN is an e-function symbol. Depending on the value of FN, the following cases arise:

  - (ZEROP FN) holds. Then FN represents the e-function symbol corresponding to IF, and EXP has the form

    (IF testpart thenpart elsepart).

    This must be evaluated in a special way. If the evaluation of the e-expression testpart yields (BTM), then the value returned is (BTM); the thenpart and elsepart e-expressions are not evaluated. If the value of testpart is NCODE-EQUAL to 0 (which serves the purpose of F here), then EV returns the result of evaluating the elsepart. Otherwise, EV returns the result of evaluating the thenpart.

  - FN is 10. Then FN represents the e-function symbol corresponding to QUOTE and EV returns the first and only argument (CAR (CDR EXP)) *without* evaluating it.

  - (SUBRP FN) is T. The definition of SUBRP is shown below. It simply checks if FN is a number between 1 and 9 (inclusive of 1 and 9). If so, FN is a primitive function symbol. In this case (and in the next case), the arguments ARGS are evaluated by EV (with FLG set to 1) and if any of the arguments evaluate to (BTM), then (BTM) is returned as the result of the evaluation. Otherwise, the result of the evaluation is obtained by applying APPLY-SUBR to FN and the values of the arguments.

  - (SUBRP FN) is F. In this case FN is a *defined* (i.e., not a primitive) e-function symbol. This is the only case where the argument N is actually used in the definition. If (ZEROP N) holds, then the bound has been exhausted and EV returns (BTM). Otherwise, the definition for the e-function FN, namely (GET (SUB1-N FN 11) FA) is recursively evaluated with the variable environment VA bound to the values of the given argument e-expressions. The expression (SUB1-N FN 11) returns the difference of FN and 11.[3]

```
Definition.
  (SUB1-N EXP N)
  =
(IF (ZEROP N)
    EXP
    (SUB1 (SUB1-N EXP (SUB1 N))))
```

```
Definition.
  (SUBRP FN)
  =
(OR (EQUAL FN '1)
    (OR (EQUAL FN '2)
        (OR (EQUAL FN '3)
            (OR (EQUAL FN '4)
                (OR (EQUAL FN '5)
                    (OR (EQUAL FN '6)
                        (OR (EQUAL FN '7)
                            (OR (EQUAL FN '8)
                                (EQUAL FN '9)))))))))
```

---

[3] Of course there is a built-in DIFFERENCE operation in Lisp, but it is given an explicit definition in the form of SUB1-N to facilitate the representability of this operation in Z2.

```
Definition.
  (EV FLG EXP VA FA N)
     =
  (IF
    (EQUAL FLG 1)
    (IF (LISTP EXP)
        (IF (OR (BTMP (EV 0 (CAR EXP) VA FA N))
                (BTMP (EV 1 (CDR EXP) VA FA N)))
            (BTM)
            (CONS (EV 0 (CAR EXP) VA FA N)
                  (EV 1 (CDR EXP) VA FA N)))
        NIL)
    (IF (LISTP EXP)
        (IF (ZEROP (CAR EXP))
            (IF (BTMP (EV 0 (CAR (CDR EXP)) VA FA N))
                (BTM)
                (IF (NCODE-EQUAL (EV 0 (CAR (CDR EXP)) VA FA N)
                                 0)
                    (EV 0
                        (CAR (CDR (CDR (CDR EXP))))
                        VA FA N)
                    (EV 0
                        (CAR (CDR (CDR EXP)))
                        VA FA N)))
            (IF (EQUAL (CAR EXP) 10)
                (CAR (CDR EXP))
                (IF (BTMP (EV 1 (CDR EXP) VA FA N))
                    (BTM)
                    (IF (SUBRP (CAR EXP))
                        (APPLY-SUBR (CAR EXP)
                                    (EV 1 (CDR EXP) VA FA N))
                        (IF (ZEROP N)
                            (BTM)
                            (EV 0
                                (GET (SUB1-N (CAR EXP) 11) FA)
                                (EV 1 (CDR EXP) VA FA N)
                                FA
                                (SUB1 N))))))))
        (GET EXP VA)))
```

This concludes the description of a Lisp interpreter written in Lisp which evaluates Lisp expressions (e-expressions) represented as Lisp objects (n-expressions). The interpreter returns the value (which is also an n-expression) of such an e-expression EXP with respect to the e-variable bindings VA and e-function definitions FA. If the computation size exceeds a prespecified bound N, then EV returns a bottom object (BTM). In the next section it is shown that the interpreter EV is representable in Z2. The subsequent sections demonstrate the computability of the metatheoretic functions by means of EV.

4.3   The Gödel-Encoding

In Chapter 2, we formalized a proof checker for Z2 in Lisp by using z-expressions to represent the syntax of Z2 within Lisp. The z-expressions consisted of natural numbers representing individual Z2 variables, lists to represent function and predicate applications, the shell constructor F-NOT to construct negations, the shell constructor F-OR to construct disjunctions, and the shell constructor FORSOME to construct existential quantifications. The task now is to map distinct z-expressions to distinct sets in Z2. A "set in Z2" simply refers to any variable-free term in Z2. This mapping is done in two stages. First we map z-expressions to n-expressions in a one-to-one manner using the Lisp function GCODE. Then we map n-expressions to sets in Z2 also in a one-to-one manner using the Lisp function NCODE.

4.3.1 Mapping Z-Expressions to N-Expressions

The mapping from Z2 expressions to n-expressions is carried out by the function GCODE. This mapping is defined by recursion on the structure of the expression and encodes constructors by means of numbers. The mapping of a CONS starts with 1, that of F-NOT with 2, F-OR with 3, and FORSOME with 4. The natural numbers are left untouched, but an object of any remaining shell is mapped to the list (CONS 5 0). In this last case, the distinctness of the mapping is violated, but all such objects are treated as NIL by the equality predicate EQL for Z2 expressions defined in page 41.

```
Definition.
  (GCODE X)
    =
  (IF (LISTP X)
      (CONS 1
            (CONS (GCODE (CAR X))
                  (GCODE (CDR X))))
      (IF (F-NOTP X)
          (CONS 2 (CONS (GCODE (ARG X)) 0))
          (IF (F-ORP X)
              (CONS 3
                    (CONS (GCODE (ARG1 X))
                          (GCODE (ARG2 X))))
              (IF (FORSOMEP X)
                  (CONS 4
                        (CONS (BIND X) (GCODE (BODY X))))
                  (IF (NUMBERP X) X (CONS 5 0)))))))
```

There is an inverse mapping corresponding to GCODE called INV-CODE which returns the expression corresponding to a given GCODE. Since all non-z-expressions are coerced to a single value in the last case of the definition of GCODE, INV-CODE is not an exact inverse, but the following statement can be proved.

```
Theorem.  INV-CODE-GCODE (rewrite):
  (EQUAL (GCODE (INV-CODE (GCODE X)))
         (GCODE X))
```

4.3.2 Mapping N-Expressions to Sets

We now show how an n-expression returned by GCODE is mapped to a set in Z2. This mapping is straightforward. In Chapter 2, we defined the numerals in Z2. Numerals are either $\emptyset$ or have the form $S(n)$, where n is a numeral. The ordered pair containing x and y is written as

$\langle x, y \rangle$ and is constructed using the pairing operation of Z2 as $\{\{x\}, \{x, y\}\}$. The Z2 expression representing an ordered pair is constructed by the function Z-OPAIR.

```
Definition.
   (Z-OPAIR X Y)
    =
   (Z-PAIR (Z-SING X) (Z-PAIR X Y))
```

Since n-expressions are built from natural numbers using the CONS operation, this mapping takes numbers to the corresponding numeral, and the mapping of a CONS is the ordered pair of the mappings of the arguments to CONS. The function NUMERAL returns the set-theoretic numeral, $S^X(\emptyset)$, corresponding to a number X.

```
Definition.
   (NUMERAL X)
    =
   (IF (ZEROP X)
       (PHI)
       (Z-SUCC (NUMERAL (SUB1 X))))
```

The function NCODE employs NUMERAL and Z-OPAIR (which constructs the ordered-pair z-expression) to return the Gödel-encoding corresponding to a given n-expression. We denote (NCODE X) as $\lceil X \rceil$. Note that $\lceil X \rceil$ is a set, that is, a variable-free term in Z2.

```
Definition.
   (NCODE X)
    =
   (IF (LISTP X)
       (Z-OPAIR (NCODE (CAR X))
                (NCODE (CDR X)))
       (NUMERAL X))
```

It is obvious that identical n-expressions yield identical Gödel-encodings. It is more difficult to show that two distinct expressions yield different Gödel-encodings, and the following two lemmas assert this to be the case. Recall that NCODE-EQUAL (defined on page 94) is the equality predicate for n-expressions. The first lemma asserts that X and Y are equal n-expressions if and only if their mappings as n-expressions are identical. The second lemma says of n-expressions X and Y that if $X \neq Y$, then $\vdash \lceil X \rceil \neq \lceil Y \rceil$. The constraints Z-SUCC-HYPS and Z-INT-HYPS on DEFNS and SYMBOLS are there to ensure that the definitions of the successor operation S and the finite-ordinal predicate I are available.

```
Theorem.  EQL-GCODE-EQUAL (rewrite):
   (EQUAL (EQL X Y)
          (EQUAL (GCODE X) (GCODE Y)))
```

```
Theorem.  NCODE-NEQUAL-PROOF-PROVES (rewrite):
   (IMPLIES (AND (SUBSET (LIST (FN 3 1)
                               (P 2 1)
                               (P 1 2)
                               (FN 2 2)
                               (FN 1 2)
                               (FN 0 0))
                         SYMBOLS)
                 (Z-SUCC-HYPS DEFNS SYMBOLS)
                 (Z-INT-HYPS DEFNS SYMBOLS)
                 (NOT (NCODE-EQUAL X Y)))
            (PROVES (NCODE-NEQUAL-PROOF X Y)
                    (F-NOT (F-EQUAL (NCODE X) (NCODE Y)))
                    GIVEN DEFNS SYMBOLS))
```

To construct proofs of inequalities in Z2, it is necessary to develop a small amount of set theory. Some of the important theorems about finite ordinals (natural numbers) and ordered pairs are listed below. The Lisp versions of these lemmas contain hypotheses asserting that certain function and predicate symbols appear in the parameter SYMBOLS and that the appropriate definitions are given in the parameter DEFNS. In the remainder of this chapter and in the next chapter, we switch conventions so that distinctly labelled variables are in fact distinct so that x, y, z, u, v, val, $trace$, etc., name different variables in Z2.

Lemma 4.3.1 $\vdash (\langle x, y \rangle = \langle u, v \rangle) \iff ((x = u) \land (y = v))$.

The right-to-left direction of the above lemma follows from the function equality axiom of Z2. The left-to-right direction is encapsulated in the two lemmas below.

```
Theorem.  Z-CAR-UNIQUE-PROOF-PROVES (rewrite):
   (IMPLIES
        (AND (TERMP (CONS X
                          (CONS Y (CONS U (CONS V NIL))))
                    1
                    SYMBOLS)
             (SUBSET (CONS (P 1 2)
                           (CONS (FN 1 2) NIL))
                     SYMBOLS))
        (PROVES (Z-CAR-UNIQUE-PROOF X Y U V)
                (F-IMPLIES (F-EQUAL (Z-OPAIR X Y)
                                    (Z-OPAIR U V))
                           (F-EQUAL X U))
                GIVEN DEFNS SYMBOLS))
```

```
Theorem.  Z-CDR-UNIQUE-PROOF-PROVES (rewrite):
   (IMPLIES
       (AND (TERMP X O SYMBOLS)
            (AND (TERMP U O SYMBOLS)
                 (AND (TERMP Y O SYMBOLS)
                      (AND (TERMP V O SYMBOLS)
                           (AND (MEMBER (P 1 2) SYMBOLS)
                                (MEMBER (FN 1 2) SYMBOLS))))))
       (PROVES (Z-CDR-UNIQUE-PROOF X U Y V)
               (F-IMPLIES (F-EQUAL (Z-OPAIR X U) (Z-OPAIR Y V))
                          (F-EQUAL U V))
               GIVEN DEFNS SYMBOLS))
```

Lemma 4.3.2 $\vdash I(n)$, *for any numeral n. (Recall that $I(x)$ is defined to ensure that x is transitive and linearly ordered by $\in$.)*

The above lemma is formally stated in the form below.

```
Theorem.  Z-INT-NUMERAL-PROOF-PROVES (rewrite):
    (IMPLIES
     (AND
      (Z-SUCC-HYPS DEFNS SYMBOLS)
      (AND
       (SUBSET
        (CONS
            (FN 3 1)
            (CONS (FN 2 2)
                  (CONS (FN 1 2)
                        (CONS (P 2 1)
                              (CONS (P 1 2)
                                    (CONS (FN 0 0) NIL))))))
        SYMBOLS)
       (Z-INT-HYPS DEFNS SYMBOLS)))
     (PROVES (Z-INT-NUMERAL-PROOF X)
             (Z-INT (NUMERAL X))
             GIVEN DEFNS SYMBOLS))
```

Lemma 4.3.3 $\vdash I(x) \supset (I(y) \supset ((S(x) = S(y)) \supset (x = y)))$.

By Lemma 4.3.3, the proof that $S(S(S(\emptyset)))$ is not equal to $S(\emptyset)$ is reduced to showing that $S(S(\emptyset))$ is not equal to $\emptyset$ which is proved using the extensionality axiom. The formal statement of the above lemma uses Z-TRANS instead of Z-INT.

```
Theorem.  Z-SUCC-EQUAL-PROOF-PROVES (rewrite):
   (IMPLIES
    (AND
     (TERMP (CONS X (CONS Y NIL)) 1 SYMBOLS)
     (AND
      (VAR-SET (CONS V1 (CONS V2 NIL)) 2)
      (AND
       (NIL-INTERSECT (CONS V1 (CONS V2 NIL))
                      (COLLECT-FREE (CONS X (CONS Y NIL))
                                    1))
      (AND
       (SUBSET
            (CONS (FN 3 1)
                  (CONS (FN 2 2)
                        (CONS (FN 1 2)
                              (CONS (P 1 2)
                                    (CONS (FN 0 0) NIL)))))
            SYMBOLS)
       (Z-SUCC-HYPS DEFNS SYMBOLS)))))
    (PROVES
     (Z-SUCC-EQUAL-PROOF X Y V1 V2)
     (F-IMPLIES
            (Z-TRANS X V1 V2)
            (F-IMPLIES (Z-TRANS Y V1 V2)
                       (F-IMPLIES (F-EQUAL (Z-SUCC X) (Z-SUCC Y))
                                  (F-EQUAL X Y))))
     GIVEN DEFNS SYMBOLS))
```

Lemma 4.3.4 $\vdash I(x) \supset ((x = \emptyset) \vee (\emptyset \in x))$.

The proof of Lemma 4.3.4 is the only place in the entire mechanical verification where the induction axiom of Z2 is used. The importance of this theorem is that it can be used to distinguish numerals from ordered pairs. Since the null set cannot be an element of an ordered pair nor equal to any ordered pair, we have $\vdash \neg I(\langle x, y \rangle)$. The formalization in Lisp of this lemma is given below.

```
Theorem.  PHI-IN-INT-PROOF-PROVES (rewrite):
   (IMPLIES
    (AND (VARIABLE X)
    (AND (Z-SUCC-HYPS DEFNS SYMBOLS)
     (AND
      (SUBSET
            (CONS (FN 3 1)
                  (CONS (FN 2 2)
                        (CONS (FN 1 2)
                              (CONS (P 1 2)
                                    (CONS (FN 0 0) NIL)))))
            SYMBOLS)
     (AND (MEMBER (P 2 1) SYMBOLS)
          (EQUAL CONCL
                 (F-IMPLIES (Z-INT X)
                            (F-OR (F-EQUAL (PHI) X)
                                  (ISIN (PHI) X))))))))
    (PROVES (PHI-IN-INT-PROOF X)
            CONCL GIVEN DEFNS SYMBOLS))
```

This concludes the discussion of the Gödel-encoding of expressions as sets in Z2. The next step is to use this encoding to prove the representability of the Lisp interpreter EV in Z2.

4.4 The Representability Theorem

Recall that $\lceil X \rceil$ denotes (NCODE X). A formula A_g is said to *represent* (in Z2) an n-ary Lisp function g if A has exactly $n+1$ free variables, say $y, x_1, \ldots, x_n$, such that:

$$\text{If } (g \ X_1 \ \ldots \ X_n) = Y,$$
$$\text{then} \quad \vdash [\lceil X_1 \rceil / x_1, \ldots, \lceil X_n \rceil / x_n, \lceil Y \rceil / y] A_g.$$
$$\text{If } (g \ X_1 \ \ldots \ X_n) \neq Y,$$
$$\text{then} \quad \vdash [\lceil X_1 \rceil / x_1, \ldots, \lceil X_n \rceil / x_n, \lceil Y \rceil / y] (\neg A_g).$$

Note that the second proof obligation is equivalent to proving $([\lceil X \rceil / x] A \supset (y = \lceil (g \ X) \rceil))$ in Z2. The implicit assumption above is that the X_i and Y are n-expressions, but since the functions we wish to represent, such as EV, behave uniformly with respect to non-n-expression arguments, the X_i and Y need not be explicitly constrained to be n-expressions.

The proofs of the representability will be illustrated with those of two Lisp functions: the Lisp primitive CAR and the recursively defined function GET. The proof of the representability of EV uses the same technique as that for GET and will therefore be omitted. The proof of the representability of computable functions in the object theory is the core of any proof of Gödel's incompleteness theorem. It is also the most tedious part of the proof. For this reason, most descriptions of the proof, including the present one, give it only cursory treatment. While the details of the representability proofs below are indeed tedious, the method is fairly simple and easy to describe.

4.4.1 Representing the Lisp Primitives

We first demonstrate the representability of the simple non-recursive Lisp primitive CAR. It is then quite easy to see how the same argument can be repeated for the other Lisp primitives. The Lisp function CAR returns X if the argument is of the form (CONS X Y), and returns 0, otherwise. Lisp primitives like CAR are represented by means of a defined function symbol in Z2. The function symbol for CAR is defined so as to return $\lceil (\text{CAR X}) \rceil$ on the argument $\lceil X \rceil$.[4] Call the Z2 function symbol for CAR, f_{car}. Its defining formula employs the logical if connective, (**if** A **then** B **else** C), which is an abbreviation for $((A \wedge B) \vee (\neg A \wedge C))$. The definition below asserts that if x encodes a list, then *val* must be the first element of x, otherwise *val* is $\emptyset$.

$$A_{\text{car}}(x, val) \triangleq (\textbf{if} \quad (\exists y, z. \ (x = \langle y, z \rangle)) \tag{4.3}$$
$$\textbf{then} \quad (\exists z. \ (x = \langle val, z \rangle))$$
$$\textbf{else} \quad (val = \emptyset))$$

---

[4]As mentioned earlier, the reason for introducing Z2 function symbols corresponding to the Lisp primitives is that it is easier to compose functions than it is to compose their defining formulas. For example, it is easier to assert, "y is the successor of the successor of x," than it is to say, "there is a z such that y is the successor of z, and z is the successor of x."

Then the defining axiom for f_{car} is $A_{car}(x, f_{car}(x))$. Before the defining formula can be admitted as an axiom, it has to be demonstrated in Z2 (using only the definitions that precede f_{car}) that for any x, such a *val* exists and is unique.

Theorem 4.4.1
$$\vdash (\exists! val.\ A_{car}(x, val)).$$

Proof. The proof of the admissibility of $A_{car}(x, f_{car}(x))$ as the defining axiom for f_{car} has two parts: those of existence and uniqueness. Both these proofs employ properties of the logical if connective. One such property used below is that if variable x does not occur free in A, then

$$\vdash \qquad (\exists x.\ (\textbf{if } A \textbf{ then } B \textbf{ else } C))$$
$$\Longleftrightarrow \quad (\textbf{if } A \textbf{ then } (\exists x.\ B) \textbf{ else } (\exists x.\ C)). \qquad (4.4)$$
$$\vdash \qquad (\exists! x.\ (\textbf{if } A \textbf{ then } B \textbf{ else } C))$$
$$\Longleftrightarrow \quad (\textbf{if } A \textbf{ then } (\exists! x.\ B) \textbf{ else } (\exists! x.\ C)). \qquad (4.5)$$

Existence: The existence part of the admissibility proof requires showing $\vdash (\exists val.\ A_{car}(x, val))$. Since *val* does not occur free in the conditional of the definition of A_{car}, namely $(\exists y, z.\ (x = \langle y, z \rangle))$, by (4.4), this reduces to showing

$$\vdash (\textbf{if} \qquad (\exists y, z.\ (x = \langle y, z \rangle)) \qquad (4.6)$$
$$\textbf{then} \quad (\exists val.\ (\exists z.\ (x = \langle val, z \rangle)))$$
$$\textbf{else} \quad (\exists val.\ ()val = \emptyset))$$

By the definition of the logical-if connective, the formula (4.6) easily follows from

$$\vdash (\exists y, z.\ (x = \langle y, z \rangle)) \supset (\exists val.\ (\exists z.\ (x = \langle val, z \rangle))) \qquad (4.7)$$
$$\vdash \neg(\exists y, z.\ (x = \langle y, z \rangle)) \supset (\exists val.\ ()val = \emptyset). \qquad (4.8)$$

The formulas (4.7) and (4.8) are easily proved in Z2. A derivation of the above form can easily be iterated when the logical-if connectives are nested, which is the case with more complicated definitions.

Uniqueness: The uniqueness part of the admissibility proof involves showing

$$\vdash ((A_{car}(x, val) \wedge A_{car}(x, val')) \supset (val = val')),$$

where val' is a new variable distinct from x and *val*. By (4.5), the uniqueness of *val* in $A_{car}(x, val)$ can be derived from $(\exists! val.\ (\exists).\ z)(x = \langle val, z \rangle)$ which follows from the uniqueness of ordered pairs (Lemma 4.3.1), and $(\exists! val.\ (val = \emptyset))$ which follows from the transitivity of equality. This method for deriving uniqueness proofs can also be extended to the case when the defining formula contains nested logical-if connectives. ∎

Having admitted $A_{car}(x, f_{car}(x))$ as the definition of $f_{car}(x)$, the main property we wish to establish is that f_{car} represents the CAR operation of Lisp under our encoding of Lisp n-expressions as Z2 sets. The representability of CAR can then be stated as below.

Theorem 4.4.2

$$\vdash f_{\text{car}}(\lceil \text{X} \rceil) = \lceil (\text{CAR X}) \rceil.$$

Proof. This proof has two cases according to whether (LISTP X) holds. If (LISTP X) is T, then by the definition of NCODE, we get $\vdash (\exists y, z. (\lceil \text{X} \rceil = \langle y, z \rangle))$. From this, and the definition of A_{car}, we have $\vdash (\exists z. (\lceil \text{X} \rceil = \langle f_{\text{car}}(\lceil \text{X} \rceil), z \rangle))$. By the definition of our Gödel-encoding, when (LISTP X), then $\lceil \text{X} \rceil$ is $\langle \lceil (\text{CAR X}) \rceil, \lceil (\text{CDR X}) \rceil \rangle$. Then $\vdash f_{\text{car}}(\lceil \text{X} \rceil) = \lceil (\text{CAR X}) \rceil$ follows by Lemma 4.3.1.

When (LISTP X) is F, then $\lceil \text{X} \rceil$ is a numeral, which by Lemma 4.3.4 is not an ordered pair, so $\vdash \neg(\exists y, z. (\lceil \text{X} \rceil = \langle y, z \rangle))$. Then $\lceil (\text{CAR X}) \rceil = \lceil 0 \rceil = \emptyset$, and from the defining axiom $A_{\text{car}}(\lceil \text{X} \rceil, f_{\text{car}}(\lceil \text{X} \rceil))$ and the definition of A_{car}, we get the desired result $\vdash (f_{\text{car}}(\lceil \text{X} \rceil) = \emptyset)$. ∎

The Lisp statements of the steps in the representability of the CAR operation of Lisp in Z2 can now be presented. The definition of the logical-if connective and its crucial properties are shown below.

```
Definition.
  (F-IF X Y Z)
  =
  (F-OR (F-AND X Y) (F-AND (F-NOT X) Z))
```

The next two theorems establish $\vdash A \supset (B \supset (\textbf{if } A \textbf{ then } B \textbf{ else } C))$ and $\vdash \neg A \supset (C \supset (\textbf{if } A \textbf{ then } B \textbf{ else } C))$, respectively.

```
Theorem.  F-IF-INTRO-STEP1 (rewrite):
  (IMPLIES (AND (FORMULA X SYMBOLS)
           (AND (FORMULA Y SYMBOLS)
                (FORMULA Z SYMBOLS)))
           (PROVES (F-IF-INTRO-STEP-PROOF1 X Y Z)
                   (F-IMPLIES X
                              (F-IMPLIES Y (F-IF X Y Z)))
                   GIVEN DEFNS SYMBOLS))
```

```
Theorem.  F-IF-INTRO-STEP2 (rewrite):
  (IMPLIES (AND (FORMULA X SYMBOLS)
           (AND (FORMULA Y SYMBOLS)
                (FORMULA Z SYMBOLS)))
           (PROVES (F-IF-INTRO-STEP-PROOF2 X Y Z)
                   (F-IMPLIES (F-NOT X)
                              (F-IMPLIES Z (F-IF X Y Z)))
                   GIVEN DEFNS SYMBOLS))
```

It can also be shown that $\vdash A \supset ((\textbf{if } A \textbf{ then } B \textbf{ else } C) \supset B)$ and $\vdash \neg A \supset ((\textbf{if } A \textbf{ then } B \textbf{ else } C) \supset C)$. One direction of the equivalence (4.4) is displayed below as FORSOME-F-IF-PROOF-PROVES.

```
Theorem.  FORSOME-F-IF-PROOF-PROVES (rewrite):
  (IMPLIES
   (AND
       (VARIABLE VAR)
       (AND (FORMULA X SYMBOLS)
            (AND (FORMULA Y SYMBOLS)
                 (AND (FORMULA Z SYMBOLS)
                      (NOT (MEMBER VAR (COLLECT-FREE X 0)))))))
    (PROVES (FORSOME-F-IF-PROOF VAR X Y Z)
            (F-IMPLIES (F-IF X
                             (FORSOME VAR Y)
                             (FORSOME VAR Z))
                       (FORSOME VAR (F-IF X Y Z)))
            GIVEN DEFNS SYMBOLS))
```

In Section 2.1.10, the Z2 function symbol for the Z2 analogue of the CAR operation was
introduced but the definition was omitted. This definition is given below. It is shown to be
admissible in Z2 and to represent the CAR operation. A Z2 expression of the form $f_{car}(s)$ is
constructed using the Z-CAR operation below.

```
Definition.
  (Z-CAR X)

     =

  (CONS (FN 4 1) (CONS X NIL))
```

The definition of the Z2 analogue of the CAR operation is just (Z-CAR-DEFN 0 (Z-CAR 0) 1
2), where the Z2 variables 0, 1, and 2 respectively represent x, y, and z from the definition
in (4.3).

```
Definition.
  (Z-CAR-DEFN X Y Y1 Z)

     =

  (F-IF (FORSOME Y1
                 (FORSOME Z
                          (F-EQUAL X (Z-OPAIR Y1 Z))))
        (FORSOME Z (F-EQUAL X (Z-OPAIR Y Z)))
        (F-EQUAL Y (PHI)))
```

The existence part of the admissibility proof for f_{car} is established by the theorem below.

```
Theorem.  Z-CAR-EXIS-PROOF-PROVES (rewrite):
  (IMPLIES
       (AND (VAR-SET (CONS Y (CONS Z NIL)) 2)
            (AND (SUBSET (CONS (FN 0 0)
                               (CONS (FN 1 2) NIL))
                         SYMBOLS)
                 (AND (TERMP X 0 SYMBOLS)
                      (EQUAL CONCL
                             (FORSOME Y (Z-CAR-DEFN X Y Y Z)))))))
       (PROVES (Z-CAR-EXIS-PROOF X Y Z)
               CONCL GIVEN DEFNS SYMBOLS))
```

The uniqueness part of the proof is not explicitly proved but is a subproof in the admissibility of the definition of f_{car}. Theorem 4.4.2 is displayed in Lisp below. There are a number of trival hypotheses, but the main point is that construction of a proof of CONCL of the form $\vdash f_{car}(\lceil X \rceil) = \lceil (CAR\ X) \rceil$.

```
Theorem.  Z-CAR-NCODE-PROOF-PROVES (rewrite):
  (IMPLIES
   (AND
    (Z-SUCC-HYPS DEFNS SYMBOLS)
    (AND
     (SUBSET
      (CONS
       (FN '4 '1)
       (CONS
        (FN '3 '1)
        (CONS (FN '2 '2)
              (CONS (FN '1 '2)
                    (CONS (P '2 '1)
                          (CONS (P '1 '2)
                                (CONS (FN '0 '0) 'NIL)))))))
      SYMBOLS)
     (AND (Z-INT-HYPS DEFNS SYMBOLS)
          (AND (Z-CAR-HYPS DEFNS SYMBOLS)
               (EQUAL CONCL
                      (F-EQUAL (Z-CAR (NCODE X))
                               (NCODE (CAR X)))))))))
   (PROVES (Z-CAR-NCODE-PROOF X)
           CONCL GIVEN DEFNS SYMBOLS))
```

Function symbols corresponding to the other primitives: CDR, LISTP, CONS, ZEROP, NUMBERP, ADD1, SUB1, and EQUAL, can be introduced in a similar fashion. The defining axioms for these are similarly defined and shown to be admissible. Function symbols are also introduced for the Z2 analogues of SUBRP, APPLY-SUBR1, APPLY-SUBR2, and APPLY-SUBR used in the definition of EV. In the case of Lisp predicates like LISTP, the Z2 analogues are defined to return the Z2 numeral $\emptyset$ corresponding to F, and the Z2 numeral $S(\emptyset)$ corresponding to T. Note that these definitions have been introduced into Z2 as conservative extensions since the defining axioms have been proved to be admissible. We list the defining axioms for the Z2 analogues of each of these functions along with the relevant representability property that was proved. The existence, uniqueness, and representability proofs of the theorems below are quite similar to the corresponding proofs for CAR above.

Defining Formula 1 (f_{cdr})

$$\vdash (\text{if } (\exists y, z.\ x = \langle y, z \rangle) \text{ then } (\exists y.\ x = \langle y, f_{cdr}(x) \rangle) \text{ else } f_{cdr}(x) = \emptyset).$$

Theorem 4.4.3

$$\vdash f_{cdr}(\lceil X \rceil) = \lceil (CDR\ X) \rceil.$$

Defining Formula 2 (f_{listp})

$$\vdash (\textbf{if } (\exists y, z.\ x = \langle y, z \rangle) \textbf{ then } (\exists y.\ f_{\text{listp}}(x) = S(\emptyset)) \textbf{ else } f_{\text{listp}}(x) = \emptyset).$$

Theorem 4.4.4

$$\vdash f_{\text{listp}}(\lceil X \rceil) = \lceil (\texttt{BOOL-FIX (LISTP X)}) \rceil.$$

Defining Formula 3 (f_{cons})

$$\vdash f_{\text{cons}}(x, y) = \langle x, y \rangle.$$

Theorem 4.4.5

$$\vdash f_{\text{cons}}(\lceil X \rceil, \lceil Y \rceil) = \lceil (\texttt{CONS X Y}) \rceil.$$

Defining Formula 4 (f_{zerop})

$$\vdash \textbf{if } \quad I(x)$$
$$\textbf{then } \quad (\textbf{if } x = \emptyset \textbf{ then } f_{\text{zerop}}(x) = S(\emptyset) \textbf{ else } f_{\text{zerop}}(x) = \emptyset)$$
$$\textbf{else } \quad f_{\text{zerop}}(x) = S(\emptyset).$$

Theorem 4.4.6

$$\vdash f_{\text{zerop}}(\lceil X \rceil) = \lceil (\texttt{BOOL-FIX (ZEROP X)}) \rceil.$$

Defining Formula 5 (f_{numberp})

$$\vdash (\textbf{if } I(x) \textbf{ then } f_{\text{numberp}}(x) = S(\emptyset) \textbf{ else } f_{\text{numberp}}(x) = \emptyset).$$

Theorem 4.4.7

$$\vdash f_{\text{numberp}}(\lceil X \rceil) = \lceil (\texttt{BOOL-FIX (NUMBERP X)}) \rceil.$$

Defining Formula 6 (f_{add1})

$$\vdash (\textbf{if } (\exists y, z.\ x = \langle y, z \rangle) \textbf{ then } f_{\text{add1}}(x) = S(\emptyset) \textbf{ else } f_{\text{add1}}(x) = S(x)).$$

Theorem 4.4.8

$$\vdash f_{\text{add1}}(\lceil X \rceil) = \lceil (\texttt{ADD1 X}) \rceil.$$

Defining Formula 7 (f_{sub1})

$$\vdash \textbf{if } \quad I(x)$$
$$\textbf{then } \quad (\textbf{if } \quad (\exists z.\ I(z) \wedge S(z) = x)$$
$$\textbf{then } \quad S(f_{\text{sub1}}(x)) = x$$
$$\textbf{else } \quad f_{\text{sub1}}(x) = \emptyset)$$
$$\textbf{else } \quad f_{\text{sub1}}(x) = \emptyset.$$

Theorem 4.4.9

$$\vdash f_{\text{sub1}}(\lceil X \rceil) = \lceil (\texttt{SUB1 X}) \rceil.$$

Defining Formula 8 (f_{equal})

$$\vdash (\textbf{if } x = y \textbf{ then } f_{\text{equal}}(x,y) = S(\emptyset) \textbf{ else } f_{\text{equal}}(x,y) = \emptyset).$$

Theorem 4.4.10

$$\vdash f_{\text{equal}}(\lceil X \rceil, \lceil Y \rceil) = \lceil (\texttt{BOOL-FIX (EQUAL X Y)}) \rceil.$$

Defining Formula 9 ($f_{\text{apply-subr1}}$)

$$\vdash \quad \textbf{if} \quad x = S^6(\emptyset)$$

$\textbf{then} \quad f_{\text{apply-subr1}}(x,y) = f_{\text{listp}}(f_{\text{car}}(y))$

$\textbf{else} \quad \textbf{if} \quad x = S^7(\emptyset)$

$\qquad \textbf{then} \quad f_{\text{apply-subr1}}(x,y) = f_{\text{cons}}(f_{\text{car}}(y), f_{\text{car}}(f_{\text{cdr}}(y)))$

$\qquad \textbf{else} \quad \textbf{if} \quad x = S^8(\emptyset)$

$\qquad\qquad \textbf{then} \quad f_{\text{apply-subr1}}(x,y) = f_{\text{car}}(f_{\text{car}}(y))$

$\qquad\qquad \textbf{else} \quad f_{\text{apply-subr1}}(x,y) = f_{\text{cdr}}(f_{\text{car}}(y)).$

Theorem 4.4.11

$$\vdash f_{\text{apply-subr1}}(\lceil FN \rceil, \lceil ARGS \rceil) = \lceil (\texttt{APPLY-SUBR1 FN ARGS}) \rceil.$$

Defining Formula 10 ($f_{\text{apply-subr2}}$)

$$\vdash \quad \textbf{if} \quad x = S^3(\emptyset)$$

$\textbf{then} \quad f_{\text{apply-subr2}}(x,y) = f_{\text{zerop}}(f_{\text{car}}(y))$

$\textbf{else} \quad \textbf{if} \quad x = S^4(\emptyset)$

$\qquad \textbf{then} \quad f_{\text{apply-subr2}}(x,y) = f_{\text{add1}}(f_{\text{car}}(y))$

$\qquad \textbf{else} \quad \textbf{if} \quad x = S^5(\emptyset)$

$\qquad\qquad \textbf{then} \quad f_{\text{apply-subr2}}(x,y) = f_{\text{sub1}}(f_{\text{car}}(y))$

$\qquad\qquad \textbf{else} \quad f_{\text{apply-subr2}}(x,y) = f_{\text{apply-subr1}}(x,y).$

Theorem 4.4.12

$$\vdash f_{\text{apply-subr2}}(\lceil FN \rceil, \lceil ARGS \rceil) = \lceil (\texttt{APPLY-SUBR2 FN ARGS}) \rceil.$$

Defining Formula 11 ($f_{\text{apply-subr}}$)

$$\vdash \quad \textbf{if} \quad x = S(\emptyset)$$

$\textbf{then} \quad f_{\text{apply-subr}}(x,y) = f_{\text{equal}}(f_{\text{car}}(y), f_{\text{car}}(f_{\text{cdr}}(y)))$

$\textbf{else} \quad \textbf{if} \quad x = S^2(\emptyset)$

$\qquad \textbf{then} \quad f_{\text{apply-subr}}(x,y) = f_{\text{numberp}}(f_{\text{car}}(y))$

$\qquad \textbf{else} \quad f_{\text{apply-subr}}(x,y) = f_{\text{apply-subr2}}(x,y).$

Theorem 4.4.13

$$\vdash f_{\text{apply-subr}}(\lceil FN \rceil, \lceil ARGS \rceil) = \lceil (\texttt{APPLY-SUBR FN ARGS}) \rceil.$$

4.4.2 Representing the Lisp Interpreter

It is significantly more difficult to represent recursively defined Lisp functions such as GET and EV than it is to represent the Lisp primitives. The problem is that there is no direct analogue to recursion in Z2. Predicates can only be defined explicitly (i.e., without recursion). Function symbols can only be introduced in place of the quantifier in a proven assertion of the form $(\exists! x.\ A)$, where A contains only previously introduced function and predicate symbols. However in recursively defined functions, as for example in the factorial function, the existence of a value when applied to $S(n)$ depends on the existence of a value for the function at n.

To represent recursive functions in Z2, we specify the *trace* of the computation of the given function in Z2. Recall that in the case of CAR, we merely specified the characteristic property of the value returned by the function. Let $\langle x, y, z \rangle$ denote $\langle x, \langle y, z \rangle \rangle$. The set that is the trace for the simple recursive function APPEND when evaluated on X and Y would be such that:

1. It contains $\langle \lceil(\texttt{APPEND X Y})\rceil, \lceil X \rceil, \lceil Y \rceil \rangle$, and

2. Every triple $\langle val, x, y \rangle$ in the trace satisfies the Z2 analogue of the definition of APPEND. That is, if x is not an ordered pair, then $val = y$. If x is an ordered pair of the form $\langle u, v \rangle$, then val must be of the form $\langle val_1, val_2 \rangle$ so that $val_1 = u$, and $\langle val_2, v, y \rangle$ is in the trace.

We will show how the specification of such a trace can be formally stated without employing recursion. The formula representing a recursive function is the assertion that there exists a trace which satisfies the Z2 analogue of its recursive definition. It should be noted that the trace of a given computation is always finite since the computation of a function is also finite.

The details of the representability of EV are quite lengthy but the basic technique is easily illustrated using the simpler recursive function GET. The function GET takes a number X and a list of values Y, and returns the X+1st element of Y. If X is 0, GET returns (CAR Y), otherwise it is equal to (GET (SUB1 X)(CDR Y)). The definition of GET is repeated below.

```
Definition.
  (GET X Y)

   =

  (IF (ZEROP X)
      (CAR Y)
      (GET (SUB1 X) (CDR Y)))
```

4.4.2.1 Defining the Formula Representing GET

The goal is to define a Z2 formula, $A_{\texttt{get}}(x, y, val)$ containing the free variables x, y, and val, such that for any n-expressions X and Y

$$\vdash\ A_{\texttt{get}}(\lceil X \rceil, \lceil Y \rceil, \lceil(\texttt{GET X Y})\rceil),\ \text{and} \tag{4.9}$$

$$\vdash\ (A_{\texttt{get}}(\lceil X \rceil, \lceil Y \rceil, val) \supset (val = \lceil(\texttt{GET X Y})\rceil)). \tag{4.10}$$

$A_{\mathbf{get}}$ employs the function symbols that were introduced for the set-theoretic analogues of CAR, CDR, SUB1, and ZEROP, called $f_{\mathbf{car}}, f_{\mathbf{cdr}}, f_{\mathbf{sub1}}$, and $f_{\mathbf{zerop}}$, respectively. The formula $A_{\mathbf{get}}(x, y, val)$ representing GET is then defined as:

$$A_{\mathbf{get}}(x, y, val) \triangleq (\exists trace.\ A_{\mathbf{get0}}(x, y, val, trace)), \qquad (4.11)$$

where $A_{\mathbf{get0}}(x, y, val, trace)$ is defined as:

$$A_{\mathbf{get0}}(x, y, val, trace) \triangleq \left\{ \begin{array}{l} (\langle val, x, y \rangle \in trace) \\ \wedge \quad (\forall x', y', val'.\ A_{\mathbf{get2}}(x', y', val', trace)). \end{array} \right. \qquad (4.12)$$

$A_{\mathbf{get2}}(x', y', val', trace)$ asserts that every element of $trace$ satisfies the analogue of the definition of GET, and is defined as:

$$
\begin{aligned}
A_{\mathbf{get2}}(x', y', val', trace) &\triangleq \\
&(\langle val', x', y' \rangle \in trace) \\
\supset \quad \mathbf{if} \quad & f_{\mathbf{zerop}}(x') = \lceil 1 \rceil \\
\mathbf{then} \quad & val' = f_{\mathbf{car}}(y') \\
\mathbf{else} \quad & \langle val', f_{\mathbf{sub1}}(x'), f_{\mathbf{cdr}}(y') \rangle \in trace.
\end{aligned}
\qquad (4.13)
$$

Therefore $A_{\mathbf{get}}(x, y, val)$ asserts the existence of a set $trace$ such that:

1. $\langle val, x, y \rangle$ belongs to $trace$, and

2. Any 3-tuple $\langle val', x', y' \rangle$ in $trace$ satisfies the analogue of the definition of GET:

    (a) If x' encodes 0, then val' is $f_{\mathbf{car}}(y')$.

    (b) Otherwise, the 3-tuple $\langle val', f_{\mathbf{cdr}}(x'), f_{\mathbf{cdr}}(y') \rangle$ corresponding to the recursive call is in the trace.

Note that in the definition of $A_{\mathbf{get}}$ above, recursion has been replaced by membership in $trace$. Definitions (4.11), (4.12), and (4.13) of $A_{\mathbf{get}}$ are used to establish the formulas (4.9) and (4.10).

4.4.2.2   Constructing the Trace for GET

Formula (4.9) is established by constructing a $trace$ corresponding to any given evaluation of GET. A Lisp object corresponding to the trace is first constructed by the function GET-GRAPH which has a recursive definition similar to that of GET. The result of applying GET-GRAPH is then Gödel-encoded as a set in Z2 by the function GRAPH-TRANS defined below. The function GET-GRAPH constructs a list of triples of the form (CONS (GET X Y) (CONS X Y)).

```
Definition.
  (GET-GRAPH X Y)

    =

  (IF (ZEROP X)
      (LIST (CONS (GET X Y) (CONS X Y)))
      (CONS (CONS (GET X Y) (CONS X Y))
            (GET-GRAPH (SUB1 X) (CDR Y))))
```

The Lisp object thus generated is converted into the required Z2 set by the function GRAPH-TRANS defined below. If X is of the form (CONS Y Z), then (GRAPH-TRANS X) is the union of {⌜Y⌝} and (GRAPH-TRANS Z). Otherwise, (GRAPH-TRANS X) is the empty set ∅. So for each element Z in (GET-GRAPH X Y), the set (GRAPH-TRANS (GET-GRAPH X Y)) contains ⌜Z⌝ as an element. We represent (GRAPH-TRANS (GET-GRAPH X Y)) as $[\![X, Y]\!]_{get}$, and note that it denotes a set in Z2.

```
Definition.
   (GRAPH-TRANS X)
       =
   (IF (LISTP X)
       (Z-UNION (Z-SING (NCODE (CAR X)))
                (GRAPH-TRANS (CDR X)))
       (PHI))
```

4.4.2.3   The Representability of GET

We now discharge the proof obligations (4.9) and (4.10). Formula (4.9) easily follows from the following theorem.

Theorem 4.4.14
$$\vdash A_{get0}(\ulcorner X \urcorner, \ulcorner Y \urcorner, \ulcorner (GET\ X\ Y) \urcorner, [\![X, Y]\!]_{get}).$$ (4.14)

Proof. By the induction corresponding to the recursion displayed in the definition of GET.[5]

Base case: Here   (ZEROP X) is T. We have by the definition of GET-GRAPH and GRAPH-TRANS, that $[\![X, Y]\!]_{get}$ is just the singleton set $\{\langle \ulcorner (GET\ X\ Y) \urcorner, \ulcorner X \urcorner, \ulcorner Y \urcorner \rangle\}$. Clearly,

$$\vdash \langle \ulcorner (GET\ X\ Y) \urcorner, \ulcorner X \urcorner, \ulcorner Y \urcorner \rangle \in [\![X, Y]\!]_{get}.$$

This establishes the first part of the definition (4.12) of A_{get0}. It is easy to show the second part of the definition, $(\forall x', y', val'.\ A_{get2}(x', y', val', [\![X, Y]\!]_{get}))$, since the triple $\langle \ulcorner (GET\ X\ Y) \urcorner, \ulcorner X \urcorner, \ulcorner Y \urcorner \rangle$ is the only element of the trace $[\![X, Y]\!]_{get}$. By the assumption that (ZEROP X) is T, we have by Theorem 4.4.6 that

$$\vdash (f_{zerop}(\ulcorner X \urcorner) = \ulcorner 1 \urcorner).$$ (4.15)

By the definition of GET and Theorem 4.4.2, we have

$$\vdash (\ulcorner (GET\ X\ Y) \urcorner = f_{car}(\ulcorner Y \urcorner)).$$ (4.16)

The formulas (4.15) and (4.16) easily yield

$$\vdash (\forall x', y', val'.\ A_{get2}(x', y', val', [\![X, Y]\!]_{get}))$$

by propositional reasoning (using the tautology theorem).

---
[5]These proofs employ induction in the metatheory and not in Z2.

Induction case: Here (ZEROP X) is F so that $[\![X, Y]\!]_{\mathbf{get}}$ is defined to be

$$\{\langle \ulcorner(\text{GET } X \text{ } Y)\urcorner, \ulcorner X\urcorner, \ulcorner Y\urcorner\rangle\} \cup [\![(\text{SUB1 } X), (\text{CDR } Y)]\!]_{\mathbf{get}}.$$

It is therefore easy to obtain the first part of the definition (4.12) of $A_{\mathbf{get0}}$:

$$\vdash \langle \ulcorner(\text{GET } X \text{ } Y)\urcorner, \ulcorner X\urcorner, \ulcorner Y\urcorner\rangle \in [\![X, Y]\!]_{\mathbf{get}}.$$

The remaining goal is to establish

$$\vdash (\forall x', y', val'. \; A_{\mathbf{get2}}(x', y', val', [\![X, Y]\!]_{\mathbf{get}})).$$

By the definitions of $[\![X, Y]\!]_{\mathbf{get}}$ and $A_{\mathbf{get2}}$, the universal quantifier $\forall$ can be distributed over the union operation $\cup$ of Z2, to yield the two subgoals:

$$\vdash (\forall x', y', val'. \; A_{\mathbf{get2}}(x', y', val', [\![(\text{SUB1 } X), (\text{CDR } Y)]\!]_{\mathbf{get}})), \text{ and} \qquad (4.17)$$

$$\begin{aligned}
\vdash \;\; &\textbf{if} \quad f_{\text{zerop}}(\ulcorner X\urcorner) = \ulcorner 1\urcorner \\
&\textbf{then} \quad \ulcorner(\text{GET } X \text{ } Y)\urcorner = f_{\text{car}}(\ulcorner Y\urcorner) \\
&\textbf{else} \quad \{\langle \ulcorner(\text{GET } X \text{ } Y)\urcorner, f_{\text{sub1}}(\ulcorner X\urcorner), f_{\text{cdr}}(\ulcorner Y\urcorner)\rangle\} \in [\![X, Y]\!]_{\mathbf{get}}.
\end{aligned} \qquad (4.18)$$

Formula (4.17) easily follows from the induction hypothesis:

$$\vdash A_{\mathbf{get0}}(\ulcorner(\text{SUB1 } X)\urcorner, \ulcorner(\text{CDR } Y)\urcorner, \ulcorner(\text{GET } X \text{ } Y)\urcorner, [\![(\text{SUB1 } X), (\text{CDR } Y)]\!]_{\mathbf{get}}) \qquad (4.19)$$

by expanding the definition of $A_{\mathbf{get0}}$.

Since (ZEROP X) is F, we have by Theorem 4.4.6 that $\vdash f_{\mathbf{zerop}}(\ulcorner X\urcorner) = \emptyset$, and hence

$$\vdash \neg(f_{\mathbf{zerop}}(\ulcorner X\urcorner) = \ulcorner 1\urcorner). \qquad (4.20)$$

Formula (4.18) can be established by propositional reasoning from

$$\vdash \{\langle \ulcorner(\text{GET } X \text{ } Y)\urcorner, f_{\text{sub1}}(\ulcorner X\urcorner), f_{\text{cdr}}(\ulcorner Y\urcorner)\rangle\} \in [\![X, Y]\!]_{\mathbf{get}}. \qquad (4.21)$$

By the definition of GET, this reduces to

$$\vdash \{\langle \ulcorner(\text{GET } (\text{SUB1 } X) \text{ } (\text{CDR } Y))\urcorner, f_{\text{sub1}}(\ulcorner X\urcorner), f_{\text{cdr}}(\ulcorner Y\urcorner)\rangle\} \in [\![X, Y]\!]_{\mathbf{get}}. \qquad (4.22)$$

Formula (4.22) easily follows from Theorems 4.4.9 and 4.4.3, the induction hypothesis (Formula (4.19)), and the definition of $A_{\mathbf{get0}}$. ∎

The above proof shows that an appropriate trace does exist (and can be constructed), and hence proves Formula (4.9). Formula (4.10) is also proved by a similar induction as shown below.

Theorem 4.4.15

$$\vdash (A_{\mathbf{get}}(\ulcorner X\urcorner, \ulcorner Y\urcorner, val) \supset (val = \ulcorner(\text{GET } X \text{ } Y)\urcorner)).$$

Proof.   We will assume the existence of a set *trace* satisfying $A_{\mathbf{get0}}(\ulcorner X\urcorner, \ulcorner Y\urcorner, val, trace)$ and show that $val = \ulcorner(\text{GET } X \text{ } Y)\urcorner$ follows. The proof employs an induction scheme similar to the recursion scheme used in the definition of GET.

Base case: Here (ZEROP X) is T. By Theorem 4.4.6 and the definitions of A_{get0} and A_{get2} (once again substituting $\lceil X \rceil$ for x', $\lceil Y \rceil$ for y', and val for val'), we get $val = f_{car}(\lceil Y \rceil)$. From this, $val = \lceil (\text{GET X Y}) \rceil$ easily follows by the definition of GET and Theorem 4.4.2.

Induction case: Here (ZEROP X) is F. By Theorem 4.4.6 and the definitions of A_{get0} and A_{get2} (substituting $\lceil X \rceil$ for x', $\lceil Y \rceil$ for y', and val for val'), we get

$$\vdash (\langle val, f_{cdr}(\lceil X \rceil), f_{cdr}(\lceil Y \rceil)\rangle \in trace). \tag{4.23}$$

The induction hypothesis has the form

$$A_{get}(\lceil (\text{SUB1 X}) \rceil, \lceil (\text{CDR Y}) \rceil, val) \supset val = \lceil (\text{GET (SUB1 X) (CDR Y)}) \rceil.$$

By the definitions (4.11) and (4.12) of A_{get} and A_{get0}, the antecedent of the induction hypothesis has the form

$$(\exists trace'. (\langle val, f_{cdr}(\lceil X \rceil), f_{cdr}(\lceil Y \rceil)\rangle \in trace) \wedge A_{get2}(trace')).$$

Let $trace$ instantiate the existential quantifier $trace'$, then the antecedent to the induction hypothesis easily follows from the assumption $A_{get0}(\lceil X \rceil, \lceil Y \rceil, val, trace)$ and Formula (4.23). By the definition of GET, the induction hypothesis then easily yields the desired conclusion

$$\vdash val = \lceil (\text{GET X Y}) \rceil.$$

∎

Therefore, $A_{get}(\lceil X \rceil, \lceil Y \rceil, val)$ is provable in Z2, when val encodes (GET X Y), and disprovable in Z2 when val encodes something other than (GET X Y).

We now present the Lisp versions of the key steps in the representability of GET in Z2. The right-hand side of the implication in the body of the definition of A_{get2} in (4.13) is defined below as A-GET-DEFN-PART.

```
Definition.
  (A-GET-DEFN-PART X Y VAL TRACE)
        =
  (F-IF (F-EQUAL (Z-ZEROP X) (NCODE 1))
        (F-EQUAL VAL (Z-CAR Y))
        (ISIN (Z-LIST (CONS VAL
                       (CONS (Z-SUB1 X)
                             (CONS (Z-CDR Y) NIL))))
              TRACE))
```

The main part of Theorem 4.4.14 is displayed below as A-GET-DEFN-OK-PROOF-PROVES. It establishes $\vdash A_{get2}(x', y', val', [\![X, Y]\!]_{get})$.

```
Theorem.  A-GET-DEFN-OK-PROOF-PROVES (rewrite):
  (IMPLIES
   (AND
    (SUBSET
     (CONS
      (FN 10 1)
      (CONS
       (FN 9 1)
       (CONS
        (FN 5 1)
        (CONS
         (FN 4 1)
         (CONS
          (FN 3 1)
          (CONS
              (FN 2 2)
              (CONS (FN 1 2)
                    (CONS (P 1 2)
                          (CONS (P 2 1)
                                (CONS (FN 0 0) NIL)))))))))))
     SYMBOLS)
    (AND
     (Z-SUCC-HYPS DEFNS SYMBOLS)
     (AND
      (Z-SUB1-HYPS DEFNS SYMBOLS)
      (AND
       (Z-ZEROP-HYPS DEFNS SYMBOLS)
       (AND
        (Z-CAR-HYPS DEFNS SYMBOLS)
        (AND (Z-CDR-HYPS DEFNS SYMBOLS)
         (AND (Z-INT-HYPS DEFNS SYMBOLS)
              (VAR-SET (CONS VAL1 (CONS X1 (CONS Y1 NIL)))
                       3)))))))))
   (PROVES
    (A-GET-DEFN-OK-PROOF X1 Y1 VAL1 X Y)
    (F-IMPLIES
            (ISIN (Z-LIST (CONS VAL1 (CONS X1 (CONS Y1 NIL))))
                  (GRAPH-TRANS (GET-GRAPH X Y)))
            (A-GET-DEFN-PART X1 Y1 VAL1
                             (GRAPH-TRANS (GET-GRAPH X Y))))
    GIVEN DEFNS SYMBOLS))
```

The definition of A_{get} is displayed below as **A-GET**. Here LIST-ALL-QUANTIFY performs the operation of universal quantification with respect to a list of variables.

```
Definition.
  (A-GET X Y VAL X1 Y1 VAL1 TRACE)
  =
  (F-AND
   (ISIN (Z-OPAIR VAL (Z-OPAIR X Y))
         TRACE)
   (LIST-ALL-QUANTIFY
    (CONS X1 (CONS Y1 (CONS VAL1 NIL)))
    (F-IMPLIES
            (ISIN (Z-LIST (CONS VAL1 (CONS X1 (CONS Y1 NIL))))
                  TRACE)
            (A-GET-DEFN-PART X1 Y1 VAL1 TRACE))))
```

The proof of Theorem 4.4.14 is now stated as shown below.

```
Theorem.  A-GET-OK-PROOF-PROVES (rewrite):
  (IMPLIES
   (AND
    (SUBSET
     (CONS
      (FN 10 1)
      (CONS
       (FN 9 1)
       (CONS
        (FN 5 1)
        (CONS
         (FN 4 1)
         (CONS
          (FN 3 1)
          (CONS
            (FN 2 2)
            (CONS (FN 1 2)
                  (CONS (P 1 2)
                        (CONS (P 2 1)
                              (CONS (FN 0 0) NIL)))))))))))
      SYMBOLS)
    (AND
     (Z-SUCC-HYPS DEFNS SYMBOLS)
     (AND
      (Z-ZEROP-HYPS DEFNS SYMBOLS)
      (AND
       (Z-SUB1-HYPS DEFNS SYMBOLS)
       (AND
        (Z-CAR-HYPS DEFNS SYMBOLS)
        (AND
         (Z-CDR-HYPS DEFNS SYMBOLS)
         (AND
          (Z-INT-HYPS DEFNS SYMBOLS)
          (AND
           (VAR-SET (CONS VAL1 (CONS X1 (CONS Y1 NIL)))
                    3)
           (AND
            (EQUAL X-BAR (NCODE X))
            (AND
             (EQUAL Y-BAR (NCODE Y))
             (AND
              (EQUAL VAL-BAR (NCODE (GET X Y)))
              (EQUAL TRACE
                     (GRAPH-TRANS (GET-GRAPH X Y)))))))))))))))
    (PROVES (A-GET-OK-PROOF1 X Y X1 Y1 VAL1)
            (A-GET X-BAR Y-BAR VAL-BAR X1 Y1 VAL1 TRACE)
            GIVEN DEFNS SYMBOLS))
```

Theorem 4.4.15 is formalized below as A-GET-GOOD-PROOF-PROVES. The correspondence between the theorem as stated above and the Lisp version below is straightforward.

```
Theorem.  A-GET-GOOD-PROOF-PROVES (rewrite):
   (IMPLIES
    (AND
     (SUBSET
      (CONS
       (FN 10 1)
       (CONS
        (FN 9 1)
        (CONS
         (FN 5 1)
         (CONS
          (FN 4 1)
          (CONS
           (FN 3 1)
           (CONS
              (FN 2 2)
              (CONS (FN 1 2)
                    (CONS (P 1 2)
                          (CONS (P 2 1)
                                (CONS (FN 0 0) NIL)))))))))))
      SYMBOLS)
     (AND
      (Z-SUCC-HYPS DEFNS SYMBOLS)
      (AND
       (Z-ZEROP-HYPS DEFNS SYMBOLS)
       (AND
        (Z-SUB1-HYPS DEFNS SYMBOLS)
        (AND
         (Z-CAR-HYPS DEFNS SYMBOLS)
         (AND
          (Z-CDR-HYPS DEFNS SYMBOLS)
          (AND
           (Z-INT-HYPS DEFNS SYMBOLS)
           (AND
            (TERMP VAL 0 SYMBOLS)
            (VAR-SET (CONS X1
                          (CONS Y1
                                (CONS VAL1 (CONS TRACE NIL))))
                  4)))))))))
     (PROVES (A-GET-GOOD-PROOF X Y VAL X1 Y1 VAL1 TRACE)
             (F-IMPLIES (A-GET (NCODE X)
                               (NCODE Y)
                               VAL X1 Y1 VAL1 TRACE)
                        (F-EQUAL VAL (NCODE (GET X Y))))
             GIVEN DEFNS SYMBOLS))
```

4.4.2.4   The Representability of EV

By means of proofs similar to those of Theorems 4.4.14 and 4.4.15, we can establish the representability in Z2 of SUB1-N and EV. In the case of EV, a formula A_{ev} can be constructed which represents the function EV, and the following theorem can be proved.

Theorem 4.4.16 *If* (EV FLG EXP VA FA N) $\neq$ (BTM), *then*

$$\vdash \ A_{ev}(\ulcorner FLG \urcorner, \ulcorner EXP \urcorner, \ulcorner VA \urcorner, \ulcorner FA \urcorner, \ulcorner (EV\ FLG\ EXP\ VA\ FA\ N) \urcorner) \qquad (4.24)$$

$$\vdash \quad \begin{aligned} &A_{ev}(\ulcorner FLG \urcorner, \ulcorner EXP \urcorner, \ulcorner VA \urcorner, \ulcorner FA \urcorner, val) \\ &\supset (val = \ulcorner (\text{EV FLG EXP VA FA N}) \urcorner). \end{aligned} \tag{4.25}$$

From (4.25), by substituting $\ulcorner VAL \urcorner$ for val, and equality and propositional reasoning, we have:

$$\text{If VAL} \neq (\text{EV FLG EXP VA FA N}), \tag{4.26}$$

$$\text{then} \quad \vdash \neg A_{ev}(\ulcorner FLG \urcorner, \ulcorner EXP \urcorner, \ulcorner VA \urcorner, \ulcorner FA \urcorner, \ulcorner VAL \urcorner).$$

We have described an encoding of n-expressions as sets, and sketched how a Lisp interpreter EV can be represented (under the above encoding) by a formula in Z2. The Lisp analogues of the above theorems can now be presented. The Lisp function A-EV-DEFN-PART1 constructs the Z2 analogue of the definition of EV. Its definition is extremely long and will be omitted. It is analogous to the consequent of the implication in the definition of A_{get2}. The EV analogue of A_{get0} is constructed by the Lisp function A-EV.

```
Definition.
  (A-EV FLG EXP VA FA VAL TRACE)
    =
  (F-AND
   (ISIN (Z-LIST (CONS VAL
                       (CONS FLG
                              (CONS EXP
                                    (CONS VA (CONS FA NIL)))))))
         TRACE)
   (LIST-ALL-QUANTIFY
                 '(6 7 8 9 10)
                 (F-IMPLIES (ISIN (Z-LIST '(10 6 7 8 9))
                                  TRACE)
                            (A-EV-DEFN-PART1 6
                                             7
                                             8
                                             9
                                             10
                                             TRACE))))
```

One part of the definition of A-EV is captured by the Lisp function A-EV-DEFN below. It is not used immediately below but is used in the construction of the undecidable sentence in the following chapter.

```
Definition.
  (A-EV-DEFN TRACE)
    =
  (LIST-ALL-QUANTIFY
                 '(6 7 8 9 10)
                 (F-IMPLIES (ISIN (Z-LIST '(10 6 7 8 9))
                                  TRACE)
                            (A-EV-DEFN-PART1 6
                                             7
                                             8
                                             9
                                             10
                                             TRACE)))
```

The representability of EV is then established by the two theorems A-EV-OK-PROOF-PROVES and A-EV-GOOD-PROOF-PROVES.

```
Theorem.  A-EV-OK-PROOF-PROVES (rewrite):
  (IMPLIES
   (AND
    (EV-HYPS GIVEN DEFNS SYMBOLS)
    (AND
        (NOT (BTMP (EV FLG EXP VA FA N)))
        (EQUAL CONCL
              (A-EV (NCODE FLG)
                    (NCODE EXP)
                    (NCODE VA)
                    (NCODE FA)
                    (NCODE (EV FLG EXP VA FA N))
                    (GRAPH-TRANS
                       (EV-GRAPH FLG EXP VA FA N))))))
   (PROVES (A-EV-OK-PROOF FLG EXP VA FA N)
          CONCL GIVEN DEFNS SYMBOLS))
```

```
Theorem.  A-EV-GOOD-PROOF-PROVES (rewrite):
  (IMPLIES
      (AND (EV-HYPS GIVEN DEFNS SYMBOLS)
          (AND (VARIABLE TRACE)
              (AND (NOT (MEMBER TRACE
                             '(6 7 8 9 10 11 12 15 16)))
                  (NOT (BTMP (EV FLG EXP VA FA N))))))
      (PROVES (A-EV-GOOD-PROOF FLG EXP VA FA N TRACE)
            (F-IMPLIES (A-EV (NCODE FLG)
                            (NCODE EXP)
                            (NCODE VA)
                            (NCODE FA)
                            10
                            TRACE)
                      (F-EQUAL 10
                              (NCODE
                                  (EV FLG EXP VA FA N))))
            GIVEN DEFNS SYMBOLS))
```

This concludes the demonstration that the Lisp interpreter EV is representable in Z2. We can now use this fact to demonstrate the representability of the metatheory of Z2 within Z2 as is needed for the construction of the undecidable sentence.

4.5   Computability of the Metatheory

While we have seen that Lisp functions such as EV are representable in Z2, we still have not demonstrated the representability of the metatheoretic functions used to define a proof checker for Z2 in Chapter 2. To show the representability of these functions, we show that to each Lisp function used in Chapter 2, it is possible to define an e-function with the same behavior when evaluated by EV. This correspondence will be made more precise, but first we note that the metatheoretic functions involve operations on z-expressions which contain constructors like F-NOT, F-OR, and FORSOME, whereas the e-expressions when evaluated by EV only operate on or return n-expressions. The function GCODE is used to encode z-expressions as n-expressions.

A function like APPEND is said to be *computable* by EV if

- There is an e-function denoted by f-append defined in the list of e-function definitions (FA)[6], and

- There is a bound N such that, for any VA:

$$\text{(EV 0 (LIST f-append X1 Y1) VA (FA) N)} = \text{(GCODE (APPEND X Y))},$$

  where

  - (EV 0 X1 VA (FA) N) is (GCODE X), and
  - (EV 0 Y1 VA (FA) N) is (GCODE Y).

The e-function symbol f-append is actually 11, the first defined e-function symbol (see definition of EV). (G-APPEND X Y) returns an e-expression with 11 as the function symbol, and X and Y as its two arguments.

```
Definition.
  (G-APPEND X Y)
    =
  (LIST 11 X Y)
```

The actual definition of the e-function corresponding to APPEND is a bit complicated due to the fact that its arguments are n-expression that encode z-expressions (via GCODE). The Lisp function PR-IF is used to construct an if-then-else e-expression.

```
Definition.
  (PR-IF X Y Z)
    =
  (CONS 0
        (CONS X (CONS Y (CONS Z NIL))))
```

The definition of the e-function analogue of APPEND is essentially an if-then-else constructed by PR-IF, where

- (G-TYPEP 0 1) constructs an e-expression that checks if the first argument (the e-variable numbered 0) is the encoding (via GCODE) of a z-expression of the form (CONS X Y).

- G-CONS, G-CAR, and G-CDR construct the e-expressions which perform the GCODE analogues of CONS, CAR, and CDR, respectively. For example, the evaluation by EV of the e-expression constructed by (G-CONS X Y) returns (GCODE (CONS x y)), where x and y are the results of evaluating X and Y, respectively.

---

[6]Here FA is a Lisp function of zero arguments. It simply names a Lisp data structure that is a list of suitable e-function definitions. The definition of FA is not very interesting and is omitted.

This definition of `f-append` and the definitions of other e-functions encoding the metatheoretic functions from Chapter 2 are included in the list constructed by (FA).

```
Definition.
  (G-APPEND-DEFN)
    =
  (PR-IF (G-TYPEP 0 1)
         (G-CONS (G-CAR 0)
                 (G-APPEND (G-CDR 0) 1))
         1)
```

The Lisp function APPEND has been shown to be admissible in the Boyer–Moore logic so its evaluation on any given arguments always terminates. On the other hand, an e-expression of the form (f-append X Y) when evaluated by EV only returns a well-defined (i.e., non-(BTM)) value when EV is given a large enough number for the argument N. A function G-APPENDN is therefore needed to bridge the gap between f-append and APPEND. If the e-expressions X and Y evaluate to x and y, respectively, then the result of evaluating the e-expression (f-append X Y), and N as the counter, is (G-APPENDN x y N). The Lisp functions PR-IFN, G-CONSN, G-CARN, and G-CDRN are similar analogues of IF, CONS, CAR, and CDR, respectively. The Lisp expression (G-TYPEPN X 1) is analogous to (LISTP X): it checks whether X evaluates to an n-expression that (via GCODE) encodes a z-expression of the form (CONS U V). The correspondence between f-append and G-APPENDN is proved as EVAL-G-APPEND below.

```
Definition.
  (G-APPENDN X Y N)
    =
  (IF (OR (ZEROP N) (BTMP X) (BTMP Y))
      (BTM)
      (PR-IFN (G-TYPEPN X 1)
              (G-CONSN (G-CARN X)
                       (G-APPENDN (G-CDRN X) Y (SUB1 N)))
              Y))
```

```
Theorem.  EVAL-G-APPEND (rewrite):
  (EQUAL (EV 0 (G-APPEND X Y) VA (FA) N)
         (G-APPENDN (EV 0 X VA (FA) N)
                    (EV 0 Y VA (FA) N)
                    N))
```

The proof of EVAL-G-APPEND employs a tricky induction which corresponds to the recursion scheme employed by the definition of G-APPEND-IND below.

```
Definition.
  (G-APPEND-IND X Y VA N)
    =
  (IF (ZEROP N)
      T
      (G-APPEND-IND (G-CDR 0)
                    1
                    (EV 1 (LIST X Y) VA (FA) N)
                    (SUB1 N)))
```

With `EVAL-G-APPEND`, the value of `f-append` has been expressed in terms of `G-APPENDN`. The lemma `G-APPENDN-GCODE` shows how `G-APPENDN` can be expressed in terms of `APPEND`. This requires that we find a suitably large N so that `G-APPENDN` does not return `(BTM)`. In the case of `APPEND`, this N must exceed `(LENGTH X)`, the length of the first of argument.

```
Theorem.  G-APPENDN-GCODE (rewrite):
   (IMPLIES (LESSP (LENGTH X) N)
            (EQUAL (G-APPENDN (GCODE X) (GCODE Y) N)
                   (GCODE (APPEND X Y))))
```

The same technique can be used to show that more complicated Lisp functions like `COLLECT-FREE`, `SUBST`, `FORMULA`, and `PROVES`, are also computable using `EV`. Lisp predicates when computed by `EV` return 0 or 1 instead of T or F. Thus in the case of `G-FORMULAN-GCODE` below, the value returned is `(BOOL-FIX (FORMULA EXP SYMBOLS))`, where `BOOL-FIX` converts T to 1, and F to 0.

```
Theorem.   G-FORMULAN-GCODE (rewrite):
    (IMPLIES (LESSP (FORMULA-CNT EXP SYMBOLS) N)
             (EQUAL (G-FORMULAN (GCODE EXP)
                                (GCODE SYMBOLS)
                                N)
                    (BOOL-FIX (FORMULA EXP SYMBOLS))))
```

In general, if `p` is some binary Lisp predicate which is defined by the e-function `f-p` in `(FA)`, then if the computation of `(p X Y)` terminates, there is an n such that for any N not smaller than n:

$$\text{(EV 0 (LIST f-p 0 1) (LIST (GCODE X)(GCODE Y)) (FA) N)}$$
$$=$$
$$\text{(BOOL-FIX (P X Y))}.$$

We can then use Formulas (4.24) and (4.27), instantiating for `FLG`, `EXP`, `VA`, `FA`, and `N`, to get the following theorem.

Theorem 4.5.1 *If* `(EV 0 (LIST f-p 0 1) (LIST (GCODE X)(GCODE Y)) (FA) N)` $\neq$ `(BTM)`, *then*

$$\vdash A_{\text{ev}} \left(\begin{array}{l} \ulcorner 0 \urcorner, \\ \ulcorner (\text{LIST f-p 0 1}) \urcorner, \\ \ulcorner (\text{LIST (GCODE X)(GCODE Y)}) \urcorner, \\ \ulcorner (\text{FA}) \urcorner, \\ \ulcorner (\text{BOOL-FIX (P X Y)}) \urcorner \end{array} \right) \qquad (4.27)$$

If `VAL` $\neq$ `(BOOL-FIX (P X Y))`, *then*

$$\vdash \neg A_{\text{ev}} \left(\begin{array}{l} \ulcorner 0 \urcorner, \\ \ulcorner (\text{LIST f-p 0 1}) \urcorner, \\ \ulcorner (\text{LIST (GCODE X)(GCODE Y)}) \urcorner, \\ \ulcorner (\text{FA}) \urcorner, \\ \ulcorner \text{VAL} \urcorner \end{array} \right) \qquad (4.28)$$

This concludes the discussion of the computability of the metatheoretic functions by means of the Lisp interpreter `EV`.

4.6 Summary

In Chapter 2 we presented a formal system consisting of the axioms and rules of Shoenfield's first-order logic and the axioms of a simple set theory due to Cohen. Proofs in this formal logic were represented as Lisp data structures, and a Lisp function PROVES was defined to check such proofs. In Chapter 3, we showed how the proof-checking capability of this simple proof checker could be extended by means of powerful derived inference rules such as the tautology theorem. The present chapter dealt mainly with the representability of Lisp functions in Z2. An encoding of Lisp data structures as sets was defined, and under this encoding, a Lisp interpreter EV was shown to be representable by a formula in Z2. The proof of the representability of EV forms a sizable part of the mechanical proof of the incompleteness theorem.

It was also argued that metatheoretic functions like FORMULA and PROVES were computable by means of EV under an encoding of Z2 expressions as n-expressions. Though it seems obvious that these functions are computable, the explicit proof of this fact had to be carried out for each of these metatheoretic functions. Informal descriptions of proofs of the incompleteness theorem often carry out this particular argument by appealing to the form of the metatheoretic functions and asserting that they are obviously recursive. However, when one is carrying out a proof in a particular theory, one can only appeal to the form of a function in that theory in its metatheory. Thus, most informal proofs of the incompleteness theorem are actually argued in the meta-metatheory of the object theory. However, as we have shown, the proof can be carried out entirely in the metatheory through the use of a Lisp interpreter written in Lisp itself. To complete the proof of the incompleteness theorem, we still need to construct an undecidable sentence in Z2.

Chapter 5

The Undecidable Sentence

We have only the deepest sympathy for those readers who have not encountered this type of simple yet mind-boggling argument before.

Marvin Minsky [Min67]

The construction of an undecidable sentence of Z2 will be completed in this chapter. This construction involves

1. Enumerating Z2 proofs using a one-to-one mapping from Z2 proofs to the natural numbers.

2. Defining a Lisp predicate that searches for the first proof or disproof (i.e., proof of the negation) of the given formula in this enumeration of proofs, and returns T or F, accordingly. Let this "theorem checker" be of the form (THEOREM X), where X is the given formula.

3. Using the representability of the above predicate to construct a sentence U which can be seen to assert, "(THEOREM u) = F," where u is the Lisp representation of U. It will be shown that if U is either provable or disprovable in Z2, then it is both provable and disprovable.

5.1   The Enumeration of Proofs

Proofs in Z2 are Lisp data structures constructed using Lisp constructors such as CONS, ADD1, F-NOT, F-OR, FORSOME. If we can enumerate all such data structures, then we can also enumerate all Z2 proofs. This enumeration is achieved by mapping these data structures (z-expressions) to n-expressions and, in turn, mapping n-expressions to natural numbers. The mapping from z-expressions to n-expressions is performed by the function GCODE discussed in the previous chapter (page 99). The resulting n-expressions are enumerated by the function NUMCODE below. If X is a number, (NUMCODE X) returns the odd number $2X+1$. If X is a list of the form (CONS Y Z), and Y and Z are mapped to y and z, respectively, then (NUMCODE X) returns $2^y \times 3^z$. Since nothing is mapped to 0, a CONS is always mapped to an even number.

```
Definition.
  (NUMCODE X)
    =
  (IF (LISTP X)
      (TIMES (EXP 2 (NUMCODE (CAR X)))
             (EXP 3 (NUMCODE (CDR X))))
      (ADD1 (DOUBLE X)))
```

Corresponding to NUMCODE, there is an inverse mapping DECIPHER so that (DECIPHER (NUMCODE X)) is X, provided X is an n-expression. When Y is an odd number, (DECIPHER Y) returns ((Y-1)/2). Let U and V be the number of factors of 2 and 3, respectively, in Y. Then if Y is an even number, (DECIPHER Y) returns (CONS u v), where u is (DECIPHER U), and v is (DECIPHER V).

In the enumeration above, the number corresponding to a Z2 proof X is (NUMCODE (GCODE X)). Given such a number Y, we can recover the proof (or one that is EQL to it) as (INV-CODE (DECIPHER Y)). The desired relationship is the one proved below.

```
Theorem.  INV-CODE-DECIPHER-NUMCODE (rewrite):
  (EQUAL (GCODE (INV-CODE (DECIPHER (NUMCODE (GCODE X)))))
         (GCODE X))
```

This concludes the first step of the construction of the undecidable sentence, namely, the enumeration of proofs. Next we construct the Lisp predicate that searches for the first proof or disproof of a given formula, in this enumeration.

5.2   The Theorem Checker

We define a Lisp predicate that searches along the enumeration of proofs until a Z2 proof of either the given formula or its negation is found. The Lisp predicate ISTHM below starts a counter M and checks if (INV-CODE (DECIPHER M)) is the required proof or disproof of X. If it is a proof of X, ISTHM returns T. If it is a disproof, a proof of the negation of X, ISTHM returns F. Otherwise, it increments M by one and repeats the process. (ISTHM X 0 GIVEN DEFNS SYMBOLS) starts the count from 0.

```
(ISTHM M X GIVEN DEFNS SYMBOLS)
=
(IF (PROVES (INV-CODE (DECIPHER M)) X GIVEN DEFNS SYMBOLS)
    T
    (IF (PROVES (INV-CODE (DECIPHER M))
                (F-NOT X) GIVEN DEFNS SYMBOLS)
        F
        (ISTHM (ADD1 M) X GIVEN DEFNS SYMBOLS)))
```

The above definition of ISTHM is not admissible under the principle of definition of the Boyer–Moore logic since it might not terminate upon execution. One can, however, define the corresponding e-function, call it f-isthm, to be evaluated by the Lisp interpreter EV (which returns (BTM) when the depth of the evaluation stack exceeds the bound N). As seen in EVAL-G-APPEND, the evaluation by EV of an e-expression starting with f-isthm can be shown to be equal to the corresponding G-ISTHMN applied to the values of the argument e-expressions. The lemmas thereafter can be stated in terms of G-ISTHMN. An e-expression of the form (f-isthm ...) is constructed by the function G-ISTHM defined below.

```
Definition.
  (G-ISTHM PFN EXP GIVEN DEFNS SYMBOLS)
    =
  (CONS 37
        (CONS PFN
              (CONS EXP
                    (CONS GIVEN
                          (CONS DEFNS (CONS SYMBOLS NIL))))))
```

The defining e-expression for the e-function f-isthm is shown below as G-ISTHM-DEFN. Note that the e-variables 0, 1, 2, 3, and 4 respectively represent the arguments M, X, GIVEN, DEFN, and SYMBOLS in the "definition" of ISTHM above. The e-expressions for the primitive QUOTE operation is constructed by PR-QUOTE, and the e-expression for ADD1 is constructed by PR-ADD1. G-PROVES, G-INV-CODE, and G-DECIPHER construct the e-function analogues of PROVES, INV-CODE, and DECIPHER, respectively.

```
Definition.
  (G-ISTHM-DEFN)
    =
  (PR-IF (G-PROVES (G-INV-CODE (G-DECIPHER 0))
                   1
                   2
                   3
                   4)
         (PR-QUOTE (BOOL-FIX T))
         (PR-IF (G-PROVES (G-INV-CODE (G-DECIPHER 0))
                          (G-F-NOT 1)
                          2
                          3
                          4)
                (PR-QUOTE (BOOL-FIX F))
                (G-ISTHM (PR-ADD1 0)
                         1
                         2
                         3
                         4)))
```

As with G-APPEND and G-APPENDN, we can define a bridge function G-ISTHMN and prove the correspondence between f-isthm and G-ISTHMN. In the definition below, (FA) is a list of all the relevant e-function definitions, but as already indicated in the footnote in page 121, the definition of FA will not be explicitly presented.

```
Theorem.  G-ISTHM-EVAL (rewrite):
   (EQUAL (EV 0
              (G-ISTHM PFN EXP GIVEN DEFNS SYMBOLS)
              VA
              (FA)
              N)
          (G-ISTHMN (EV 0 PFN VA (FA) N)
                    (EV 0 EXP VA (FA) N)
                    (EV 0 GIVEN VA (FA) N)
                    (EV 0 DEFNS VA (FA) N)
                    (EV 0 SYMBOLS VA (FA) N)
                    N))
```

When (ISTHM M X GIVEN DEFNS SYMBOLS) returns T, we can find a proof of X. This proof is (INV-CODE (DECIPHER PFN)), where PFN is the value of the first argument on the call to ISTHM on which the computation terminates. Similarly, if (ISTHM M X GIVEN DEFNS SYMBOLS) returns F, we can find a proof of (F-NOT X). If X is either provable or disprovable, then ISTHM is guaranteed to terminate, since the given proof or disproof provides an upper bound on the search along the enumeration of proofs. We can summarize these observations in the following theorems.

Theorem 5.2.1 *If* (ISTHM M X GIVEN DEFNS SYMBOLS) *is* T, *then for some* N, (PROVES (INV-CODE (DECIPHER N)) X GIVEN DEFNS SYMBOLS) *is* T. *Thus when* DEFNS *is* (THM-DEFNS) *and* SYMBOLS *is* (THM-SYMBS), *the conclusion yields* ⊢ X.

Theorem 5.2.2 *If* (ISTHM M X GIVEN DEFNS SYMBOLS) *is* F, *then for some* N, (PROVES (INV-CODE (DECIPHER N)) (F-NOT X) GIVEN DEFNS SYMBOLS) *is* T. *Again, when* DEFNS *is* (THM-DEFNS) *and* SYMBOLS *is* (THM-SYMBS), *the conclusion yields* ⊢ ¬X.

The Lisp analogues of these two theorems are displayed below. The theorem FIND-PROOF-PROVES-T-CASE asserts that when G-ISTHMN returns the GCODE encoding of T, then G-FIND-PROOF returns the GCODE encoding of a proof of EXP. Correspondingly, the theorem FIND-PROOF-PROVES-F-CASE asserts that G-FIND-PROOF constructs a proof of the GCODE encoding of the negation of EXP, when G-ISTHMN returns the GCODE encoding of F. The function G-FINDN returns the value of the counter N when the recursive evaluation of f-isthm terminated. This value of the counter is sufficient to successfully terminate the evaluation of G-PROVESN. In the definitions below, G-INV-CODEN and G-DECIPHERN are variants of INV-CODE and DECIPHER, respectively, that employ the counter argument N.

```
Definition.
  (G-FINDN PFN EXP GIVEN DEFNS SYMBOLS N)
     =
  (IF
   (OR (BTMP PFN)
       (OR (BTMP EXP)
           (OR (BTMP GIVEN)
               (OR (BTMP DEFNS)
                   (OR (BTMP SYMBOLS) (ZEROP N))))))
   (BTM)
   (PR-IFN
       (G-PROVESN (G-INV-CODEN (G-DECIPHERN PFN (SUB1 N))
                              (SUB1 N))
                   EXP GIVEN DEFNS SYMBOLS
                   (SUB1 N))
       (SUB1 N)
       (PR-IFN (G-PROVESN (G-INV-CODEN (G-DECIPHERN PFN
                                                   (SUB1 N))
                                      (SUB1 N))
                          (G-F-NOTN EXP)
                          GIVEN DEFNS SYMBOLS
                          (SUB1 N))
               (SUB1 N)
               (G-FINDN (PR-ADD1N PFN)
                        EXP GIVEN DEFNS SYMBOLS
                        (SUB1 N)))))
```

```
Definition.
  (G-FIND-PROOF PFN EXP GIVEN DEFNS SYMBOLS N)
     =
  (IF
   (OR (BTMP PFN)
       (OR (BTMP EXP)
           (OR (BTMP GIVEN)
               (OR (BTMP DEFNS)
                   (OR (BTMP SYMBOLS) (ZEROP N))))))
   (BTM)
   (PR-IFN
       (G-PROVESN (G-INV-CODEN (G-DECIPHERN PFN (SUB1 N))
                              (SUB1 N))
                   EXP GIVEN DEFNS SYMBOLS
                   (SUB1 N))
       (G-INV-CODEN (G-DECIPHERN PFN (SUB1 N))
                    (SUB1 N))
       (PR-IFN (G-PROVESN (G-INV-CODEN (G-DECIPHERN PFN
                                                   (SUB1 N))
                                      (SUB1 N))
                          (G-F-NOTN EXP)
                          GIVEN DEFNS SYMBOLS
                          (SUB1 N))
               (G-INV-CODEN (G-DECIPHERN PFN (SUB1 N))
                            (SUB1 N))
               (G-FIND-PROOF (PR-ADD1N PFN)
                             EXP GIVEN DEFNS SYMBOLS
                             (SUB1 N)))))
```

```
    Theorem.  FIND-PROOF-PROVES-T-CASE
  (rewrite):
     (IMPLIES (EQUAL (G-ISTHMN PFN EXP GIVEN DEFNS SYMBOLS N)
                     (BOOL-FIX T))
              (EQUAL (G-PROVESN (G-FIND-PROOF PFN EXP GIVEN DEFNS
                                             SYMBOLS N)
                                EXP GIVEN DEFNS SYMBOLS
                                (G-FINDN PFN EXP GIVEN DEFNS
                                         SYMBOLS N))
                     (BOOL-FIX T)))
```

```
    Theorem.  FIND-PROOF-PROVES-F-CASE (rewrite):
     (IMPLIES (EQUAL (G-ISTHMN PFN EXP GIVEN DEFNS SYMBOLS N)
                     (BOOL-FIX F))
              (AND (EQUAL (G-PROVESN (G-FIND-PROOF PFN EXP GIVEN
                                                  DEFNS SYMBOLS N)
                                     EXP GIVEN DEFNS SYMBOLS
                                     (G-FINDN PFN EXP GIVEN DEFNS
                                              SYMBOLS N))
                          (BOOL-FIX F))
                   (EQUAL (G-PROVESN (G-FIND-PROOF PFN EXP GIVEN
                                                  DEFNS SYMBOLS N)
                                     (G-F-NOTN EXP)
                                     GIVEN DEFNS SYMBOLS
                                     (G-FINDN PFN EXP GIVEN DEFNS
                                              SYMBOLS N))
                          (BOOL-FIX T))))
```

The above theorems can now be recast in terms of PROVES rather than G-PROVESN. The theorems below also replace EXP by (SUBST EXP VAR TERM FLG) in preparation for the construction of the undecidable sentence in the next section. The function FIND-PROOFN returns the PFN such that (PROVES (INV-CODE (DECIPHER PFN)) ...) is T. The function THM-COUNT ensures that the counter value at the point where the evaluation of G-ISTHMN terminates is large enough to successfully evaluate G-PROVESN, G-INV-CODEN, and G-DECIPHERN.

```
Theorem.  FIND-PROOF-PROVES-T-CASE-GCODE (rewrite):
  (IMPLIES
   (AND
    (LESSP
     (ADD1
         (THM-COUNT (FIND-PROOFN PFN
                                 (GCODE
                                   (SUBST EXP VAR TERM FLG))
                                 (GCODE GIVEN)
                                 (GCODE DEFNS)
                                 (GCODE SYMBOLS)
                                 N)
                    EXP VAR TERM FLG GIVEN DEFNS SYMBOLS))
     (G-FINDN (GCODE PFN)
              (GCODE (SUBST EXP VAR TERM FLG))
              (GCODE GIVEN)
              (GCODE DEFNS)
              (GCODE SYMBOLS)
              N))
    (EQUAL (G-ISTHMN (GCODE PFN)
                     (GCODE (SUBST EXP VAR TERM FLG))
                     (GCODE GIVEN)
                     (GCODE DEFNS)
                     (GCODE SYMBOLS)
                     N)
           (BOOL-FIX T)))
   (PROVES
    (INV-CODE
          (DECIPHER (FIND-PROOFN PFN
                                 (GCODE
                                   (SUBST EXP VAR TERM FLG))
                                 (GCODE GIVEN)
                                 (GCODE DEFNS)
                                 (GCODE SYMBOLS)
                                 N)))
    (SUBST EXP VAR TERM FLG)
    GIVEN DEFNS SYMBOLS))
```

```
Theorem.  FIND-PROOF-PROVES-F-CASE-GCODE (rewrite):
   (IMPLIES
    (AND
     (LESSP
      (ADD1
          (THM-COUNT (FIND-PROOFN PFN
                                  (GCODE (SUBST EXP VAR TERM 0))
                                  (GCODE GIVEN)
                                  (GCODE DEFNS)
                                  (GCODE SYMBOLS)
                                  N)
                     (F-NOT EXP)
                     VAR TERM 0
                     GIVEN DEFNS SYMBOLS))
      (G-FINDN (GCODE PFN)
               (GCODE (SUBST EXP VAR TERM 0))
               (GCODE GIVEN)
               (GCODE DEFNS)
               (GCODE SYMBOLS)
               N))
     (EQUAL (G-ISTHMN (GCODE PFN)
                      (GCODE (SUBST EXP VAR TERM 0))
                      (GCODE GIVEN)
                      (GCODE DEFNS)
                      (GCODE SYMBOLS)
                      N)
            (BOOL-FIX F)))
    (PROVES
     (INV-CODE
          (DECIPHER (FIND-PROOFN PFN
                                 (GCODE (SUBST EXP VAR TERM 0))
                                 (GCODE GIVEN)
                                 (GCODE DEFNS)
                                 (GCODE SYMBOLS)
                                 N)))
     (F-NOT (SUBST EXP VAR TERM 0))
     GIVEN DEFNS SYMBOLS))
```

The next step is to use the computability of the above metatheoretic functions (including ISTHM) by EV, and the representability of EV, to construct a sentence which asserts that a proof of its own negation occurs first in the enumeration of proofs.

5.3   Construction of the Undecidable Sentence

The only remaining task is to actually construct the undecidable sentence. We first define a Lisp predicate on one variable that we then represent in Z2 (using the representability theorem) by means of a formula with one free variable. The undecidable sentence turns out to be a particular substitution instance of this formula. The construction of the undecidable sentence is parametric in the additional axioms assumed in GIVEN so that we can show Z2 to be essentially incomplete.

Let THEOREM be defined as follows. be

$$(\text{THEOREM X GIVEN}) \triangleq (\text{ISTHM 0 X GIVEN (THM-DEFNS)(THM-SYMBS)}) \qquad (5.1)$$

The function THEOREM is also only used for the informal exposition here and does not appear in the mechanical proof. As with ISTHM, if the given formula X is either provable or disprovable, then the computation of (THEOREM X GIVEN) is guaranteed to terminate returning T or F.

Theorem 5.3.1 *If* X *is bound to a Z2 formula that is either provable or disprovable in Z2, then the evaluation of the Lisp expression* (THEOREM X GIVEN) *terminates with the value* T *or the value* F.

Let the Lisp predicate P now be defined as:

$$(P\ A\ GIVEN) \triangleq \qquad (5.2)$$
$$(THEOREM\ (SUBST\ A\ 6\ (NCODE\ (GCODE\ A))\ 0)\ GIVEN).$$

In the above definition, the second argument to SUBST is the object-theoretic variable 6 which is labelled y for the purpose of the informal exposition below. The reason GIVEN is a parameter to P is that the construction of the undecidable sentence is itself parametric with respect to GIVEN. The undecidable sentence in the statement of the incompleteness theorem presented in Chapter 2 was of the form (UNDECIDABLE-SENTENCE GIVEN). So, for any specific GIVEN, there will be a fixed undecidable sentence. If (P A GIVEN) is T, then the result of substituting the set (NCODE (GCODE A)) (abbreviated as $\lceil$(GCODE A)$\rceil$) in Z2 for the variable y in A, has a proof in the enumeration of proofs which occurs before any disproof of it. Let A be the Lisp representation of the Z2 formula A, then we have by Theorem 5.2.1 and the definitions of P and THEOREM, that

$$\text{If } (P\ A\ GIVEN) = T, \text{ then } \vdash [\lceil(GCODE\ A)\rceil/y]A. \qquad (5.3)$$

When (P A GIVEN) is F, it asserts that the disproof of the result of substituting the term $\lceil$(GCODE A)$\rceil$ for y in A, occurs prior to any proof of the same, in the enumeration. So, we have by Theorem 5.2.2, and the definitions of P and THEOREM, that

$$\text{If } (P\ A\ GIVEN) = F, \text{ then } \vdash \neg([\lceil(GCODE\ A)\rceil/y]A). \qquad (5.4)$$

Whenever the computation of the Lisp predicate P terminates, there is a sufficiently large N, so that

$$(EV\ 0\ (LIST\ f\text{-}p\ 0\ 1)\ (LIST\ (GCODE\ A)(GCODE\ GIVEN))\ (FA)\ N) \\ = (BOOL\text{-}FIX\ (P\ A\ GIVEN)) \qquad (5.5)$$

Now consider the Z2 formula which represents (by Formulas (4.27) and (4.28)), the left-hand side of the Equality (5.5) above, when its value is VAL:

$$A_{ev}\left(\begin{array}{l} \lceil 0\rceil, \\ \lceil(LIST\ f\text{-}p\ 0\ 1)\rceil, \\ \lceil(LIST\ (GCODE\ A)(GCODE\ GIVEN))\rceil, \\ \lceil(FA)\rceil, \\ \lceil VAL\rceil \end{array}\right) \qquad (5.6)$$

Formula (5.6) contains no free Z2 variables and is therefore a sentence. Now let us set the value VAL in Formula (5.6) to (BOOL-FIX F) to get

$$
A_{\mathrm{ev}} \left(
\begin{array}{l}
\lceil 0 \rceil, \\
\lceil (\text{LIST f-p 0 1}) \rceil, \\
\lceil (\text{LIST (GCODE A)(GCODE GIVEN))} \rceil, \\
\lceil (\text{FA}) \rceil, \\
\lceil (\text{BOOL-FIX F}) \rceil
\end{array}
\right)
\tag{5.7}
$$

The resulting sentence (5.7) is provable if (P A GIVEN) is F, and disprovable if (P A GIVEN) is T (by (4.27) and (4.28), respectively, and Equation (5.5)). By the definition of NCODE (namely, $\lceil . \rceil$), we can replace

$$
\lceil (\text{LIST (GCODE A)(GCODE GIVEN))} \rceil
$$

in the above Z2 formula (5.7) by

$$
\langle \lceil (\text{GCODE A}) \rceil, \lceil (\text{LIST (GCODE GIVEN))} \rceil \rangle,
$$

to get

$$
A_{\mathrm{ev}} \left(
\begin{array}{l}
\lceil 0 \rceil, \\
\lceil (\text{LIST f-p 0 1}) \rceil, \\
\langle \lceil (\text{GCODE A}) \rceil, \lceil (\text{LIST (GCODE GIVEN))} \rceil \rangle, \\
\lceil (\text{FA}) \rceil, \\
\lceil (\text{BOOL-FIX F}) \rceil
\end{array}
\right)
\tag{5.8}
$$

The Z2 sentence (5.8) above can be obtained from the Z2 formula G defined below by substituting the term $\lceil (\text{GCODE A}) \rceil$ for the free variable y. The sentence (5.8) can then be abbreviated as $[\lceil (\text{GCODE A}) \rceil / y]G$.

$$
G \triangleq A_{\mathrm{ev}} \left(
\begin{array}{l}
\lceil 0 \rceil, \\
\lceil (\text{LIST f-p 0 1}) \rceil, \\
\langle y, \lceil (\text{LIST (GCODE GIVEN))} \rceil \rangle, \\
\lceil (\text{FA}) \rceil, \\
\lceil (\text{BOOL-FIX F}) \rceil
\end{array}
\right)
\tag{5.9}
$$

Note that the construction of the formula G is parametric in in the metatheoretic variable GIVEN so its Lisp representation can be written as (G GIVEN). Since y is the only free Z2 variable in G, there are no free variables left in the result of the substitution, $[\lceil (\text{GCODE A}) \rceil / y]G$. We have from (4.27) and (5.5)

$$
\text{If (P A GIVEN)} = \text{F, then} \vdash [\lceil (\text{GCODE A}) \rceil / y]G.
\tag{5.10}
$$

Similarly, from (4.28) and (5.5), we have

$$
\text{If (P A GIVEN)} = \text{T, then} \vdash \neg([\lceil (\text{GCODE A}) \rceil / y]G).
\tag{5.11}
$$

The undecidable sentence (UNDECIDABLE-SENTENCE GIVEN) used in the statement of the incompleteness theorem (page 69) is defined to be

$$
(\text{SUBST (G GIVEN) 6 (NCODE (GCODE (G GIVEN))) 0)}
$$

which is the same as $[\lceil(\text{GCODE } (\text{G GIVEN}))\rceil/y]G$. Note that the undecidable sentence defined above indeed does not contain any free variables. It is abbreviated below as U.

Theorem 5.3.2 (incompleteness) *If either $\vdash U$ or $\vdash \neg U$, then both $\vdash U$ and $\vdash \neg U$.*

Proof. The undecidable sentence U is just

$$(\text{SUBST } (\text{G GIVEN}) \ 6 \ (\text{NCODE } (\text{GCODE } (\text{G GIVEN}))) \ 0).$$

By the definition of P we then have that (P (G GIVEN) GIVEN) is just

$$(\text{THEOREM } (\text{SUBST } (\text{G GIVEN}) \ 6 \ (\text{NCODE } (\text{GCODE } (\text{G GIVEN}))) \ 0) \ \text{GIVEN}).$$

Since U is assumed to be provable or disprovable, then by Theorem 5.3.1, the evaluation of (P (G GIVEN) GIVEN) terminates yielding either T of F.

If (P (G GIVEN) GIVEN) is T, then by (5.3) (with G for A),

$$\vdash [\lceil(\text{GCODE } (\text{G GIVEN}))\rceil/y]G. \tag{5.12}$$

By (5.11) (with G for A), we also get

$$\vdash \neg([\lceil(\text{GCODE } (\text{G GIVEN}))\rceil/y]G). \tag{5.13}$$

Since U is both provable and disprovable, it follows that Z2 is inconsistent.

On the other hand, if (P (G GIVEN) GIVEN) is F, then by (5.10) (with (G GIVEN) for A),

$$\vdash [\lceil(\text{GCODE } (\text{G GIVEN}))\rceil/y]G, \tag{5.14}$$

and by (5.4) (with (G GIVEN) for A),

$$\vdash \neg([\lceil(\text{GCODE } (\text{G GIVEN}))\rceil/y]G) \tag{5.15}$$

Once again, since U is both provable and disprovable, it follows that Z2 is inconsistent. ∎

Hence, either the undecidable sentence has no proof or disproof and *the theory is incomplete, or the theory is inconsistent.* This corresponds to the statement of the incompleteness theorem presented in Section 2.2.

Note that the above construction of the undecidable sentence is parametric in a list of axioms GIVEN. Adding any particular undecidable sentence as a new axiom in GIVEN only leads to a different undecidable sentence. This means that Z2 is *essentially incomplete* in the sense that it remains incomplete under the addition of finitely many new axioms.

Intuitively, the undecidable sentence, label it U, can be interpreted as asserting, "there exists a disproof of U in the enumeration of proofs which occurs before any proof of U." Therefore if there is such a disproof, then we can also prove U. The negation of the undecidable sentence asserts that either there is no disproof at all or it is preceded by a proof of U. Then any disproof, that is, proof of the negation of U, is preceded by a proof of U. In this case it is the negation of the undecidable sentence which (assuming consistency) is "truthful" since there is no disproof. A surprising observation here is that the negation of the undecidable sentence above is essentially the same as Rosser's form of the undecidable sentence [Ros36], but the above construction is less *ad hoc* than Rosser's construction.

The informal argument outlined above deviates from the Lisp formalization in several ways. We now describe the details of the construction of the undecidable sentence as formalized in Lisp. As already mentioned, the functions ISTHM, THEOREM and P are not really part of the machine-checked proof since they are not total functions. The machine proof does contain the e-function f-isthm described in page 127. There is no e-function corresponding to the predicate P in the mechanical proof. The e-expression equivalent to P can be derived from the e-expression constructed by the function G-UNDEC shown below. The function G-UNDECN shown below is analogous to G-APPENDN. The connection between G-UNDEC and G-UNDECN is established by the theorem G-UNDEC-EVAL. The e-function GC-NCODE is the analogue of the Lisp operation (NCODE (GCODE X)) as established by the theorems below.

```
Definition.
  (GC-NCODE X)
    =
  (CONS 38 (CONS X NIL))
```

```
Definition.
  (GC-NCODE-DEFN)
    =
  (PR-IF (PR-LISTP 0)
         (G-Z-OPAIR (GC-NCODE (PR-CAR 0))
                    (GC-NCODE (PR-CDR 0)))
         (G-NUMERAL 0))
```

```
Definition.
  (GC-NCODEN X N)
    =
  (IF (OR (BTMP X) (ZEROP N))
      (BTM)
      (PR-IFN (PR-LISTPN X)
              (G-Z-OPAIRN (GC-NCODEN (PR-CARN X) (SUB1 N))
                          (GC-NCODEN (PR-CDRN X) (SUB1 N)))
              (G-NUMERALN X (SUB1 N))))
```

```
Theorem.  GC-NCODE-EVAL (rewrite):
  (EQUAL (EV 0 (GC-NCODE X) VA (FA) N)
         (GC-NCODEN (EV 0 X VA (FA) N) N))
```

```
Theorem.  GC-NCODEN-GCODE-AGAIN (rewrite):
  (IMPLIES (LESSP (NCODE-CNT (GCODE X)) N)
           (EQUAL (GC-NCODEN (GCODE X) N)
                  (GCODE (NCODE (GCODE X)))))
```

```
Definition.
  (G-UNDEC PFN EXP VAR GIVEN DEFNS SYMBOLS)
    =
  (G-ISTHM PFN
           (G-SUBST EXP VAR
                    (GC-NCODE EXP)
                    (PR-QUOTE (GCODE 0)))
           GIVEN DEFNS SYMBOLS)
```

```
Definition.
   (G-UNDECN PFN EXP VAR GIVEN DEFNS SYMBOLS N)

     =

(G-ISTHMN PFN
           (G-SUBSTN EXP VAR
                     (GC-NCODEN EXP N)
                     (GCODE 0)
                     N)
           GIVEN DEFNS SYMBOLS N)
```

```
Theorem.  G-UNDEC-EVAL (rewrite):
   (EQUAL (EV 0
              (G-UNDEC PFN EXP VAR GIVEN DEFNS SYMBOLS)
              VA
              (FA)
              N)
          (G-UNDECN (EV 0 PFN VA (FA) N)
                    (EV 0 EXP VA (FA) N)
                    (EV 0 VAR VA (FA) N)
                    (EV 0 GIVEN VA (FA) N)
                    (EV 0 DEFNS VA (FA) N)
                    (EV 0 SYMBOLS VA (FA) N)
                    N))
```

We can now examine the details of the construction of the undecidable sentence. The z-expression constructed by A-UNDEC1 is a step in the construction of the formula G. The constructed formulas has the same form as A_{ev} and uses A-EV-DEFN (defined on page 119). Note how the Z2 variable 6 already occurs in the resulting z-expression as the only element of the encoding of the VA argument to EV. If for some N, (EV FLG EXP (LIST (GCODE V)) FA N) is not (BTM), then from the theorems A-EV-OK-PROOF-PROVES and A-EV-GOOD-PROOF-PROVES, we know that the Z2 formula (A-UNDEC1 FLG EXP FA VAL) with (NCODE (GCODE V)) substituted for the Z2 variable 6, is provable when VAL is equal to (EV FLG EXP VA FA N), and is disprovable, otherwise.

```
Definition.
   (A-UNDEC1 FLG EXP FA VAL)

     =

(FORSOME 1
 (F-AND
  (ISIN
   (Z-LIST
    (CONS (NCODE VAL)
          (CONS (NCODE FLG)
                (CONS (NCODE EXP)
                      (CONS (Z-OPAIR 6 (NCODE (GCODE NIL)))
                            (CONS (NCODE FA) NIL))))))
    1)
  (A-EV-DEFN 1)))
```

A-UNDEC1 and G-UNDEC are combined in the definition of A-UNDEC below to construct a Z2 formula. The construction of A-UNDEC is parametric GIVEN, DEFNS, and SYMBOLS. When DEFNS is instantiated with (THM-DEFNS) and SYMBOLS is instantiated with (THM-SYMBS), we get a Z2 formula that is equivalent to G. Recall that the construction of G is also parametric in GIVEN. The formula constructed by A-UNDEC with (NCODE (GCODE V)) substituted for 6 is provable when

```
(EV 0 (G-UNDEC (PR-QUOTE (GCODE 0))
                0
                (PR-QUOTE (GCODE 6))
                (PR-QUOTE (GCODE GIVEN))
                (PR-QUOTE (GCODE DEFNS))
                (PR-QUOTE (GCODE SYMBOLS)))
       (LIST V)
       FA
       N)
```

evaluates to (BOOL-FIX F), and is disprovable if the above evaluation of EV returns any other non-(BTM) value. By the definitions of G-UNDECN and G-ISTHMN, and the theorems FIND-PROOF-PROVES-F-CASE-GCODE, if the value returned by the above invocation of EV is (BOOL-FIX F), then

```
(F-NOT (SUBST V 6 (NCODE (GCODE V)) 0))
```

is provable, and otherwise, the Z2 formula

```
(SUBST V 6 (NCODE (GCODE V)) 0)
```

is provable.

```
Definition.
  (A-UNDEC GIVEN DEFNS SYMBOLS)
  =
  (A-UNDEC1 0
            (G-UNDEC (PR-QUOTE (GCODE 0))
                     0
                     (PR-QUOTE (GCODE 6))
                     (PR-QUOTE (GCODE GIVEN))
                     (PR-QUOTE (GCODE DEFNS))
                     (PR-QUOTE (GCODE SYMBOLS)))
            (FA)
            (BOOL-FIX F))
```

The undecidable sentence is (UNDEC-SENT GIVEN (THM-DEFNS) (THM-SYMBOLS)), where UNDEC-SENT is as defined below. The preceding argument shows how the theorems A-EV-OK-PROOF-PROVES, A-EV-GOOD-PROOF-PROVES, FIND-PROOF-PROVES-F-CASE-GCODE, FIND-PROOF-PROVES-T-CASE-GCODE together yield a contradiction if the undecidable sentence is either provable or disprovable.

```
Definition.
  (UNDEC-SENT GIVEN DEFNS SYMBOLS)
  =
  (SUBST (A-UNDEC GIVEN DEFNS SYMBOLS)
         6
         (NCODE (GCODE (A-UNDEC GIVEN DEFNS SYMBOLS)))
         0)
```

```
Definition.
  (UNDECIDABLE-SENTENCE GIVEN)
    =
  (UNDEC-SENT GIVEN
              (THM-DEFNS)
              (THM-SYMBS))
```

This concludes the description of the mechanical proof of the incompleteness theorem.

5.4 Analysis

In the preceding pages, we described a machine-checked proof of Gödel's incompleteness theorem. The statement that was proved asserted that the theory Z2 was either incomplete or inconsistent. The proof was checked using the Boyer–Moore theorem prover. The proof was supplied to the theorem prover as a sequence of definitions and lemmas leading to the statement and proof of the incompleteness theorem. This sequence of user-supplied definitions and lemmas spans about 400 typed pages, over 2000 definitions and lemmas, and about 20,000 lines of Lisp. This proof script is distributed along with the Boyer–Moore theorem prover.

The first part of the proof was the definition of the proof checker for Z2 proofs. The definition of this proof checker covers about 14 pages. The incompleteness theorem was stated in terms of this proof checker. It is important to note that in order to be convinced that the incompleteness theorem was correctly stated, only these definitions need be studied. The rest of the script of the proof need not be examined if one is willing to believe that the Boyer–Moore theorem prover is sound.

The second part of the proof was the derivation of new inference rules from the primitive axioms and rules provided in the proof checker PROVES. These new inference rules render formal proofs considerably easier to compose, and significantly more efficient to proof check. The tautology theorem is perhaps the most important and heavily-used derived inference rule. The definitions and lemmas in this proof span about 20 pages. The lemmas and definitions in the proofs of the soundness of various other derived inference rules involving properties of quantifiers, substitution, equality, etc., cover another 40 pages.

The third part of the proof was to define the encoding of Lisp data structures as sets. For this, numerals and ordered pairs had to be defined in Z2. A number of properties of numerals and ordered pairs also had to be proved in Z2. These definitions and lemmas cover nearly 40 pages.

The fourth and perhaps the most significant part of the mechanical proof was the proof of the representability of Lisp functions in Z2. This was achieved by representing a Lisp interpreter by a formula in Z2. The script (i.e., the definitions and lemmas) of the proofs of the representability of the Lisp primitives such as ZEROP, ADD1, SUB1, NUMBERP, EQUAL, LISTP, CONS, CAR, and CDR, covers about 55 pages. The representability of the function GET occupies about 20 pages. The script of the proof of the representability of the Lisp interpreter EV occupies about 70 pages. The large length was mainly because EV is a fairly complicated function and the definitions of some of the proof constructors needed to establish its representability spanned over 5 pages. Also, numerous derived inference rules about the logical-if connective were needed to carry out this representability proof.

The fifth part of the proof was the demonstration that all of the metatheoretic functions defined in the Boyer–Moore logic were computable by means of the Lisp interpreter EV. The data structures used to represent Z2 expressions and proofs had to be translated into n-expressions (data structures built from numbers using only CONS). The definitions and lemmas in this part of the proof take up approximately 70 pages.

The sixth part of the proof established the enumerability of Z2 proofs. To enumerate Z2 proofs, n-expressions had to be encoded as numbers, and these numbers were shown to be decipherable back into the original n-expressions. The major part of the proof of the invertibility of the encoding is a proof of the unique prime factorization theorem. This part of script occupies about 25 pages.

The seventh and final part of the mechanical proof was the construction of the undecidable sentence and the proof of the statement of the incompleteness theorem. This part of the proof spans about 25 pages.

The entire mechanical proof was the result of about eighteen months of work with the Boyer–Moore theorem prover. Prior to the work done on the Boyer–Moore theorem prover, a sketch of the proof was developed. In outline, the final mechanical proof turned out to be identical to this initial sketch. When the original sketch was worked out, it soon became apparent that the most difficult part would be the proof of the representability theorem. There was then no clear idea as to how this could be approached. It was only during the machine verification that the importance of derived inference rules became apparent. A very large number of derived inference rules were needed in order to represent EV in Z2. Though many of these rules were quite complicated, it was relatively straightforward to prove them sound using the Boyer–Moore theorem prover. It was important to formulate a "good", that is, general and frequently used, set of rules in order to make the construction of formal proofs easier.

The proof of the incompleteness theorem as originally planned was not meant to include definitional extensions to Z2. This constraint turned out to be problematic because in the representability proofs, functions are constantly composed with one another, as in (CADDDR X), and it is tedious to represent such expressions in Z2 without the corresponding function symbols.

The original intention was to have this proof carried out using the Boyer–Moore theorem prover without any modifications. This goal was not achieved. Two minor modifications were made. These modifications were actually improvements which were suggested by the mechanical proof. Both modifications have been incorporated into a new version of the theorem prover. This theorem prover has been benchmarked with a list of proofs including that of the incompleteness theorem. In principle, both these modifications could have been avoided, but at a fairly severe cost. The first modification was to strengthen the theorem prover's rewriter to recognize as true those hypotheses which are tautologously true. Such hypotheses arise during applications of the tautology theorem as a derived inference rule. The unmodified version of the theorem prover would fail to simplify such hypotheses to T and would therefore fail to use the tautology theorem. An interesting consequence of this strengthening was that the modified theorem prover was actually more efficient on the benchmark proofs. The second modification was to disallow the immediate expansion of ground terms into their values. The unmodified theorem prover would expand any variable-free term to a constant representing its value by running a Lisp interpreter on it. This means that terms like (PHI) which represent Z2 expressions would always be expanded to their full form. This would be extremely inconvenient since we would like to state rewrite rules about the null set (PHI) in terms of (PHI) rather than its underlying representation. A feature was added to the theorem prover by means of which a user could disable or enable the expansion of all ground terms except those involving only the primitive functions, which always remain enabled.

5.5   Conclusions

Gödel's incompleteness theorem represents a significant landmark in mathematics. The proof of this theorem is considered to be among the deepest and most difficult of mathematical proofs. Most textbooks do little more than sketch the outlines of the argument. Our experiment was mainly to determine if the currently available proof checking technology was powerful enough to verify a significant piece of mathematics.

The proof was carried out within a constructive axiomatization of pure Lisp. In fact the proof was carried out almost entirely within primitive recursive arithmetic (PRA). The only non-primitive recursive function used was the Lisp interpreter, EV. This function exhibits a nested recursion.[1]

One of the things that distinguishes this proof from the previous proofs, apart from the fact that it has been mechanically checked, is that only modest use has been made of the natural numbers. It is traditional in most proofs to Gödel-encode expressions of the object theory as numbers which are somehow defined in the object theory. The Gödel-encoding used here is direct and straightforward. The only crucial use of numbers is to show that if proofs can be enumerated by mapping them to the numbers in a one-to-one manner, then the theory is decidable if it is complete. Another distinguishing aspect of the proof is the use of a Lisp interpreter to formalize the notion of computability. A Lisp interpreter has the nice property of being useful as a practical, as well as theoretical, model of computation. The third feature of the mechanical proof is that the method used to prove the representability of computable functions is simple and general. Most treatments of proofs of the incompleteness theorem treat the representability of primitive recursive functions as a special case.

Metamathematics plays a crucial role in computer proof checking itself through the use of derived inference rules and metatheorems which relate theories to one another. The Boyer–Moore theorem prover is particularly well-suited for the task of mechanically checking proofs in metamathematics. The use of axioms for inductively constructed data structures, the ability to introduce recursively defined functions, and the ability to make powerful use of induction within a constructive logic, are important for proofs in metamathematics.

---

[1]Feferman [Fef82] has shown that the incompleteness proof can be carried out within PRA.

Chapter 6

A Mechanical Proof of the Church–Rosser Theorem

> *By relieving the brain of all unnecessary work,*
> *a good notation sets it free to concentrate*
> *on more advanced problems, and in effect*
> *increases the mental power of the race.*
> A. N. Whitehead [Whi58]

The Church–Rosser theorem is a central metamathematical result about the lambda calculus. This chapter presents a formalization and proof of the Church–Rosser theorem that was verified using the Boyer–Moore theorem prover.[1] The proof presented in this chapter is based on that of Tait and Martin-Löf. The mechanical proof illustrates the effective use of the Boyer–Moore theorem prover in proof-checking difficult metamathematical proofs. The syntactic representation of terms turns out to be crucial to the mechanical verification. The *form* of the formalization often significantly influences the ease or difficulty of a verification. We also compare the length of the proof input to the Boyer–Moore prover with the lengths of various informal presentations of proofs of the Church–Rosser theorem.

The lambda calculus was introduced by Church [Bar78a, Chu41] in order to study functions as rules of computation rather than as graphs of argument–value pairs. It was hoped that the lambda calculus would provide an alternative foundation for logic and mathematics. This aim has remained unfulfilled due to the appearance of contradictions when the lambda calculus was extended with logical notions. The lambda calculus has nevertheless been a fruitful medium for the study of functions and computations. The programming language in which the mechanical proof was formalized, a variant of pure Lisp [MAE+65], was one of the first languages whose design was influenced by the lambda calculus.

The Church–Rosser theorem [CR36] implies the consistency of the lambda calculus. The history behind the theorem and its proofs is very interesting since it took many attempts and over thirty years to construct a plausible and widely accepted proof of the Church–Rosser theorem. To paraphrase Rosser [Ros83]:

> The original proof of CRT [the Church–Rosser theorem] was fairly long and very complicated. . . . Newman generalized the universe of discourse . . . . He proved a

---

[1]This chapter is based on an article [Sha88a] published in the *Journal of the ACM*. Copyright 1988, Association for Computing Machinery, Inc. Revised and reprinted by permission.

result similar to CRT by topological arguments. Curry ...generalized the Newman result .... Unfortunately, it turned out that neither the Newman result nor
the Curry generalization entailed CRT. ... This was discovered by Schroer ....
Schroer derived still further generalizations of the Newman and Curry results,
which indeed do entail CRT. ...Schroer 1965 is 627 typed pages .... Chapter 4
of Curry and Feys 1958 is devoted to a proof of CRT for lambda calculus and
...is not recommended for light reading. ...Meanwhile a genuine simplification
of the proof of CRT had come in sight. See Martin-Löf 1972. It is agreed that
Martin-Löf got some of his ideas from lectures by W. Tait. An exposition of
the proof of CRT according to Tait and Martin-Löf appears in Appendix I of
Hindley, Lercher and Seldin 1972.

The mechanical proof presented in this chapter is based on the Tait–Martin-Löf proof
and is an example of a machine verification of a difficult informal argument. Since the
Tait–Martin-Löf proof is a relatively recent one and involves considerable combinatorial
case-analysis, it made an interesting candidate for mechanical verification. Part of the proof
involves the use of a representation of lambda calculus terms due to de Bruijn [dB72].
The rest of this chapter is organized as follows: Section 6.1 is a brief introduction to
the lambda calculus. In Section 6.2, the Church–Rosser theorem is stated and its proof is
informally sketched. In both Sections 6.1 and 6.2, the standard notation for the lambda
calculus will be used. Section 6.3 describes the formalization of the lambda calculus and a
statement of the Church–Rosser theorem in the Boyer–Moore logic. The same formalization
in terms of the de Bruijn representation for the lambda calculus terms is carried out in
Section 6.4. Section 6.5 covers the highlights of the mechanical proof. In it, the key lemmas
are stated without proof. Section 6.6 presents the entire output from the Boyer–Moore prover
of a proof of one of the main lemmas. Section 6.7 presents an analysis of the mechanical
proof and several observations on it.

6.1   An Introduction to the Lambda Calculus

This section presents the lambda calculus notions needed to understand the proof. The
formalization of the lambda calculus within the Boyer–Moore logic will be presented in
Section 6.3. A clear exposition of the lambda calculus and the related topic of *combinatory
logic* can be found in the book *Introduction to Combinatory Logic* by Hindley, Lercher, and
Seldin [HLS72], the more recent *Introduction to Combinators and λ-Calculus* by Hindley and
Seldin [HS86], and Stenlund's *Combinators, λ-terms and Proof Theory* [Ste72]. These books
also contain informal presentations of the proof of the Church–Rosser theorem. Barendregt's
The Lambda Calculus [Bar78a] is a comprehensive volume on the subject.

6.1.1   Terms

The formal system of the lambda calculus consists of *terms* (or lambda-terms) and certain
rules for transforming terms. Intuitively, terms in lambda calculus denote one-argument
functions. Terms are constructed starting from *constants* and *variables* (collectively labelled
atoms). Some confusion might arise from the use of the words 'constant' and 'variable',

but unless otherwise mentioned, these refer to lambda calculus constants and variables. Constants do not play any role in the proof. The metavariables a, b, and c will be used to indicate constants. A term that is a variable represents an arbitrary one-argument function. The syntactic variables x, y, and z will be used to represent variables. The actual form of the variables will not be specified except to say that they form a denumerable set and are distinguishable from the constants. The uppercase letters $A, B, C, M, N, W, X, Y,$ and Z will be used as metavariables for terms. There are two ways of forming new terms. The first of these is termed *lambda-abstraction*. If M is a term and x is a variable, then the term $(\lambda x.\ M)$ represents the lambda-abstraction of M with respect to x. This means that $(\lambda x.\ M)$ is now a one-argument function with a dummy argument x. The second way of forming terms is called an *application*. The term $(X\ Y)$ represents the application of the *function* X to the *argument* Y. The argument to a function and the resulting value can both be functions. The process of *evaluating* a function application will be discussed later.

Thus the notion of a term in lambda calculus has been defined inductively by showing how terms can be constructed starting from atoms (the base case), by the operations of lambda-abstraction and application (the inductive cases). This inductive definition provides a corresponding recursion which will be used extensively. These are definitions where the base case is defined for atoms; the recursive call for a term $(\lambda x.\ M)$ is on M; and the recursive calls for a term $(M\ N)$ are on the terms M and N. The phrase *"recursion on the structure of a term"* is used to identify a recursion of this kind. Correspondingly, many proofs involve *induction on the structure of a term*.

Examples of terms

1. a

2. $(x\ a)$

3. $(\lambda x.\ a)$

4. $((\lambda x.\ (x\ x))\ (\lambda x.\ (x\ x)))$

5. $((\lambda x.\ ((\lambda y.\ y)\ u))\ v)$

The expression $X \equiv Y$ can be read as "X is identical to Y." In situations where new metavariables are introduced in the right-hand side, '$\equiv$' is usually read as "of the form." For example, $X \equiv (\lambda x.\ M)$ is read as "X is of the form $(\lambda x.\ M)$."

A few relations on terms need to be defined before the transformations on terms can be described. X is said to be a *subterm* of Y if either $X \equiv Y$, or Y is of the form $(\lambda x.\ M)$ and X is a subterm of M, or Y is of the form $(M\ N)$ and X is a subterm of either M or N.

The phrase X *occurs in* Y is an alternative way of saying X is a subterm of Y. An *occurrence* of X in Y refers to a specific location in Y where X occurs. The term $(\lambda x.\ (x\ x))$ has two occurrences in the term $((\lambda x.\ (x\ x))\ (\lambda x.\ (x\ x)))$.

If $(\lambda x.\ M)$ is a subterm of a term Y, then all the occurrences of x in M are said to be *bound* in Y. Those variable occurrences which are not bound in Y are said to be *free* in Y. For example, y is free but x is bound in the term $(\lambda x.\ (x\ y))$. These notions will now be used to define the important operation of *substitution*.

6.1.2   Substitution

Substitution is the operation of replacing all the free occurrences of a variable in a term by another term. The result of substituting X for the variable x in Y will be denoted by $[X/x]Y$ (to be read as "X for x in Y"). The analogous operation for expressions in first-order logic has already been described in Section 2.1.3. To ensure that such a substitution has the intended meaning, no free variable occurrence in X can be allowed to become bound in $[X/x]Y$. If this holds of the given X, x, and Y, then X is said to be *free for x in Y*. As described in Section 6.1.3, it is always possible to rename the bound variables so as to ensure that a term is free for a variable in another term. Substitution is defined recursively as follows:

$$[X/x]x \;\equiv\; X$$
$$[X/x]y \;\equiv\; y, \text{ if } x \not\equiv y$$
$$[X/x]a \;\equiv\; a$$
$$[X/x](\lambda y.\ M) \;\equiv\; \begin{cases} (\lambda y.\ M), & \text{if } x \equiv y \\ (\lambda y.\ [X/x]M), & \text{otherwise} \end{cases}$$
$$[X/x](M\ N) \;\equiv\; ([X/x]M\ \ [X/x]N).$$

In the remainder of this chapter, whenever the expression $[X/x]Y$ is used, it will be assumed that X is free for x in Y, unless otherwise specified.

Examples of substitution

1. $[a/x](\lambda x.\ x) \equiv (\lambda x.\ x)$

2. $[(\lambda x.\ (x\ x))/x](xx) \equiv ((\lambda x.\ (x\ x))\ (\lambda x.\ (x\ x)))$

3. $[(\lambda x.\ y)/x](\lambda y.\ x) \equiv (\lambda y.\ (\lambda x.\ y))$ (Note: $(\lambda x.\ y)$ is not free for x in $(\lambda y.\ x)$).

6.1.3   Alpha-steps and Beta-steps

There are two rules in lambda calculus for transforming terms. Both rules will be defined as relations between a term and its transformed version. The first rule permits the renaming of bound variables and is called an α-*step*. X goes to Y in an α-step (denoted by $X \overset{\alpha}{\mapsto} Y$) iff X and Y are identical except that the names used to denote the occurrences of variables bound by a λ in X and the corresponding λ in Y might differ uniformly. This vague definition of an α-step will be made precise in Section 6.3.

Examples of Alpha-steps

1. $((\lambda x.\ x)z) \overset{\alpha}{\mapsto} ((\lambda y.\ y)\ z)$

2. $(\lambda x.\ (x\ (\lambda y.\ (x\ y)))) \overset{\alpha}{\mapsto} (\lambda y.\ (y\ (\lambda x.\ (y\ x))))$

3. $((\lambda x.\ (\lambda y.\ (x\ y)))\ (\lambda z.\ (y\ z))) \overset{\alpha}{\mapsto} ((\lambda x.\ (\lambda z.\ (x\ z)))(\lambda z.\ (y\ z)))$

An α-step can be used to rename bound variables in a term Y to ensure that a given term X is free for x in the transformed version of Y.

The *evaluation* of terms takes place through a rule known as a β-*step*. It is similar to the operation of replacing dummy parameters with actual parameters in a subprogram. A subterm of the form $((\lambda x.\ M)\ N)$ is called a *redex*. A redex $((\lambda x.\ M)\ N)$ is reduced by means of *beta-reduction* to the term that is got by replacing all the free occurrences of x in M by N, i.e., the term $[N/x]M$. A term X goes to Y in a β-step iff Y is got by replacing some non-overlapping redexes in X by their beta-reduced forms. Two subterms of a term overlap if they both contain the same occurrence of at least one subterm. So the beta-reductions in a β-step must be such that they do not affect one another. The relation 'X goes to Y in a β-step' will be denoted by $X \overset{\beta}{\mapsto} Y$.

Examples of Beta-steps

1. $((\lambda x.\ (a\ x))\ b) \overset{\beta}{\mapsto} (a\ b)$

2. $(((\lambda x.\ y)\ a)\ ((\lambda x.\ (x\ x))\ b)) \overset{\beta}{\mapsto} (y\ (b\ b))$

3. $((\lambda x.\ (x\ x))\ (\lambda x.\ (x\ x))) \overset{\beta}{\mapsto} ((\lambda x.\ (x\ x))\ (\lambda x.\ (x\ x)))$

4. $((\lambda y.\ ((\lambda x.\ x)\ (b\ y)))\ ((\lambda x.\ (x\ x))\ a)) \overset{\beta}{\mapsto} ((\lambda x.\ x)\ (b\ ((\lambda x.\ (x\ x))\ a)))$

5. $((\lambda y.\ ((\lambda x.\ x)\ (b\ y)))\ ((\lambda x.\ (x\ x))\ a)) \overset{\beta}{\mapsto} ((\lambda y.\ (b\ y))\ (a\ a))$

Note that in the examples 4 and 5 above, the same term is transformed to two different terms by the two β-steps. These two examples will be employed in the next section.

A term X *reduces to* a term Y iff Y is obtained from X by a finite (possibly empty) series of α-steps or β-steps. This means either $X \equiv Y$ or there must exist terms $Z_0,\ \ldots,\ Z_n$ such that $X \equiv Z_0$, $Y \equiv Z_n$, and for all i with $0 \le i < n$, either $Z_i \overset{\alpha}{\mapsto} Z_{i+1}$ or $Z_i \overset{\beta}{\mapsto} Z_{i+1}$. The relation X reduces to Y is denoted simply as $X \longrightarrow Y$.

In summary, the lambda calculus consists of terms and transformations on terms. Terms are either atoms, lambda-abstractions or applications. Substitution is the operation of replacing all the occurrences of a certain free variable in one term by another term. There are two rules for transforming terms: α-steps and β-steps. A sequence of α-steps and β-steps can be used to reduce one term to another. These basic lambda calculus notions will be used in the next section to state the Church–Rosser theorem and to sketch a proof of it.

6.2 A Proof-Sketch of the Church–Rosser Theorem

The outline of the proof of the Church–Rosser theorem given in this section also applies to the mechanical proof, but some of the details of the mechanical proof are different.

6.2.1 The Statement

The Church–Rosser theorem [CR36] can be stated in terms of the definitions in the previous section as follows:

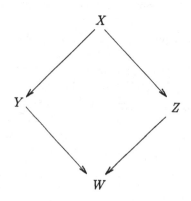

Figure 6.1: The Church–Rosser Theorem

If $X \longrightarrow Y$ and $X \longrightarrow Z$, then there exists a W such that $Y \longrightarrow W$ and $Z \longrightarrow W$.
The theorem is depicted pictorially in Figure 6.1. It can therefore be said to assert the *diamond property* of the relation '$\longrightarrow$'.

The remainder of this section provides a brief outline of the key steps in the proof of the Church–Rosser theorem. The notion of the *transitive closure* of a relation plays an important role in the proof. A relation $(X \; R \; Y)$ is the transitive closure of another relation $(U \; S \; W)$ iff the following holds of R and S:[2]

$$\forall X, Y: X R Y$$
$$\textit{iff}$$
$$\exists n, \exists Z_0, \ldots, Z_n: X \equiv Z_0 \; \textit{and}$$
$$Y \equiv Z_n \; \textit{and}$$
$$\forall i: 0 \leq i < n: Z_i S Z_{i+1}.$$

The proof of the Church–Rosser theorem consists of the following steps:

1. Showing that the transitive closure of a relation with the diamond property has the diamond property.[3]

2. Defining the relation X *walks to* Y, such that the relation $X \longrightarrow Y$ is its transitive closure.

3. Proving that the *walks to* relation has the diamond property.

Clearly if we can carry out the above three steps, we have a proof of the diamond property of the relation '$\longrightarrow$', and hence the Church–Rosser theorem.

---

[2]Since all objects in the Boyer–Moore logic are finite, there is no way to define an operation which returns the transitive closure of an arbitrary relation. Instead the transitive closure operation is defined for specific relations. Also, since there are no quantifiers in the logic, the Z_i have to be provided as an additional argument.

[3]In the mechanical proof, this lemma is proved only for a specific relation with the diamond property.

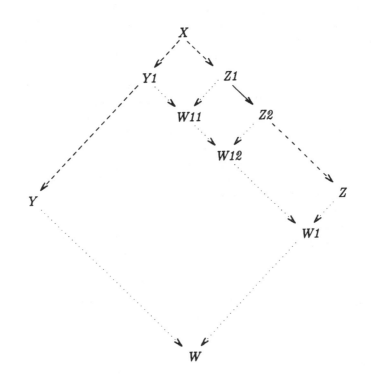

Figure 6.2: The Diamond Property over Transitive Closures

6.2.2 The Diamond Property over Transitive Closures

A diagram is employed to establish Step 1. Though this diagram is given in terms of the relations X *reduces to* Y, and X *walks to* Y, no property of these relations is used in the justification save for the fact that the former is the transitive closure of the latter. The relation X *walks to* Y is represented by $X \mapsto Y$. Since the relation $X \longrightarrow Y$ is the transitive closure of $X \mapsto Y$, it can be represented by a series of walks as $X \equiv Y_0 \mapsto Y_1 \mapsto \ldots \mapsto Y_{n-1} \mapsto Y_n \equiv Y$. Figure 6.2 shows how the diamond for relation $X \longrightarrow Y$ can be constructed from the diamonds for the relation $X \mapsto Y$. Given that $X \longrightarrow Y$ and $X \longrightarrow Z$, the figure shows how a W such that $Y \longrightarrow W$ and $Z \longrightarrow W$ can be constructed by repeatedly constructing the smaller diamonds along a row to derive W_1, W_2, and so on.

6.2.3 The Definition of Walk

The second step in the proof is to define the notion of a walk or the relation $X \mapsto Y$. This relation must have the diamond property and the relation $X \longrightarrow Y$ must be definable as its transitive closure. The relation X *goes to* Y *in a single α-step or β-step* has the second property but does not possess the diamond property. The appropriate counterexample is given below.

Counterexample: It can be constructed using nothing more than β-steps. Consider the examples 4 and 5 used to illustrate β-steps. Let X be $((\lambda y.\ ((\lambda x.\ x)\ (b\ y)))\ ((\lambda x.\ (x\ x))\ a))$, and Y be $((\lambda x.\ x)\ (b\ ((\lambda x.\ (x\ x))\ a)))$, and Z be $((\lambda y.\ (b\ y))\ (a\ a))$. As the examples show, $X \overset{\beta}{\mapsto} Y$ and $X \overset{\beta}{\mapsto} Z$. The task is to construct a W such that $Y \overset{\beta}{\mapsto} W$ and $Z \overset{\beta}{\mapsto} W$. Clearly, W is not one of Y or Z. The only other possibility is $(b\ (a\ a))$. Z goes to $(b\ (a\ a))$ in a single β-step, but Y needs two overlapping beta-reductions. Overlapping beta-reductions are not permitted in a β-step.

One point to notice in the above counterexample is that though Y could only go to $(b\ (a\ a))$ by two overlapping beta-reductions, these were such that the inner of these two reductions could be performed first. This example suggests working with a less restricted relation than a β-step in which overlapping beta-reductions are permitted, provided the inner redex is reduced before the outer one. Note that when two subterms overlap, one must always be the subterm of the other. Therefore it makes sense to talk of the reduction on an inner redex and an outer one. In any case, this less restricted form of a β-step turns out to work. This new step is called a *walk*. The rest of the proof makes the definition of a walk precise, and demonstrates that it has the diamond property.

For reasons of brevity, α-steps will be ignored in the remainder of the discussion and it will be assumed that for any redex $((\lambda x.\ M)\ N)$, N is always free for x in M.

The relation $X \mapsto Y$ ("X walks to Y") is defined recursively as follows:

1. If X is an atom, then $X \mapsto Y$ iff $Y \equiv X$.

2. If $X \equiv (\lambda x.\ M)$, then $X \mapsto Y$ iff $Y \equiv (\lambda x.\ M_Y)$ and $M \mapsto M_Y$.

3. If $X \equiv (M\ N)$, then $X \mapsto Y$ iff either:

   (a) $Y \equiv (M_Y\ N_Y)$, $M \mapsto M_Y$, and $N \mapsto N_Y$, or
   (b) $M \equiv (\lambda x.\ A)$, $M_Y \equiv (\lambda x.\ A_Y)$, $A \mapsto A_Y$, $N \mapsto N_Y$, and $Y \equiv [N_Y/x]A_Y$.

The above definition employs a recursion on the structure of the term X. Clearly, a single β-step can be represented as a walk. It is also the case that any walk can be represented as a series of β-steps. Therefore, the relation X *reduces to* Y forms the transitive closure of the relation X *walks to* Y. The above definition of a walk plays an important role in the proof of the diamond property of walks.

6.2.4   The Diamond Property of Walks

All that is needed now to complete this sketch of the proof of the Church–Rosser theorem is a proof of the diamond property of walks. The proof of the diamond property requires some crucial lemmas which we state here without proof.

Lemma 6.2.1 *If x different from y, and y not free in N, then*

$$[N/x][Z/y]M \equiv [[N/x]Z/y][N/x]M.$$

Lemma 6.2.2 (substitutivity of walks) *If $A_Y \mapsto A_W$ and $N_Y \mapsto N_W$, then* $[N_Y/x]A_Y \mapsto [N_W/x]A_W$.

The diamond property could be stated as:

Theorem 6.2.3 (diamond property of walks) *If $X \mapsto Y$ and $X \mapsto Z$, then there exists a W such that $Y \mapsto W$ and $Z \mapsto W$.*

Proof. The proof of the diamond property is by induction on the structure of the term X. Three cases arise and in each case, the appropriate W is constructed as follows:

Case 1: [X is an atom]
If X is an atom, then by the definition of a walk, $X \equiv Y \equiv Z$. Then, let W be X itself. By the definition of a walk, $Y \mapsto W$ and $Z \mapsto W$.

Case 2: [$X \equiv (\lambda x.\, M)$]
By the definition of a walk, there must exist M_Y and M_Z such that $Y \equiv (\lambda x.\, M_Y)$ and $Z \equiv (\lambda x.\, M_Z)$, so that $M \mapsto M_Y$ and $M \mapsto M_Z$. Applying the induction hypothesis to M yields M_W such that $M_Y \mapsto M_W$ and $M_Z \mapsto M_W$. Let W be $(\lambda x.\, M_W)$. Then, by the definition of a walk, both Y and Z walk to W.

Case 3: [$X \equiv (M\ N)$]
This splits up into five subcases depending on whether or not X is a redex, and if it is, whether or not it is beta-reduced in the walk. The subcases are as follows:

1. X is not a redex.

2. X is a redex and is beta-reduced in the walk to Y and not Z.

3. X is a redex and is beta-reduced in the walk to Z and not Y.

4. X is a redex and is beta-reduced in the walk to Y and to Z.

5. X is a redex but is not beta-reduced in the walk to Y or to Z.

The subcases 1 and 5 turn out to have the same proof. Subcases 2 and 3 are symmetrical, and so the proof of subcase 3 will be omitted.

Subcases 1, 5: [X is not a redex or is a redex which is not beta-reduced in the walks to Y and Z]
By the definition of a walk, there must exist M_Y, N_Y, M_Z, N_Z such that $Y \equiv (M_Y\ N_Y)$ and $Z \equiv (M_Z\ N_Z)$, so that $M \mapsto M_Y$, $N \mapsto N_Y$, $M \mapsto M_Z$, and $N \mapsto N_Z$. Applying the induction hypothesis to both M and N, we derive M_W and N_W such that

$$M_Y \ \mapsto\ M_W \tag{6.1}$$
$$M_Z \ \mapsto\ M_W \tag{6.2}$$
$$N_Y \ \mapsto\ N_W \tag{6.3}$$
$$N_Z \ \mapsto\ N_W. \tag{6.4}$$

Now if we let W be $(M_W\ N_W)$, we can conclude from (6.1), (6.2), (6.3), (6.4), and the definition of a walk that both Y and Z walk to W.

Subcase 2:   [X is a redex which is only reduced on the walk to Y]
Since X is a redex, let $X \equiv ((\lambda x.\ a)\ N)$. From the definition of a walk applied to $X \mapsto Y$ and $X \mapsto Z$, there must exist A_Y, N_Y, A_Z, N_Z such that $Y \equiv [N_Y/x]A_Y$ and $Z \equiv ((\lambda x.\ A_Z)\ N_Z)$, so that $A \mapsto A_Y$, $N \mapsto N_Y$, $A \mapsto A_Z$, and $N \mapsto N_Z$. Therefore, by the induction hypothesis applied to A and N, there must exist A_W and N_W such that

$$A_Y \ \mapsto \ A_W \tag{6.5}$$
$$A_Z \ \mapsto \ A_W \tag{6.6}$$
$$N_Y \ \mapsto \ N_W \tag{6.7}$$
$$N_Z \ \mapsto \ N_W. \tag{6.8}$$

Let W be $[N_W/x]A_W$. It can be shown that Y (of the form $[N_Y/x]A_Y$) walks to W (of the form $[N_W/x]A_W$), from (6.5), (6.7), and Lemma 6.2.2. Showing that Z (of the form $((\lambda x.\ A_Z)\ N_Z)$) walks to W involves (6.6), (6.8), and the definition of a walk.

Subcase 4:   [X is a redex and is reduced on the walks to both Y and Z]
This subcase is very similar to the previous one. As before, let $X \equiv ((\lambda x.\ a)\ N)$. The only difference is that Z is now of the form $[N_Z/x]A_Z$. The induction hypothesis yields A_W and N_W satisfying (6.5), (6.6), (6.7), and (6.8). If we set W to $[N_W/x]A_W$, then the conclusion follows from Lemma 6.2.2.

   This completes the proof of the diamond property of walks.

   The proof of Lemma 6.2.2 proceeds by a similar induction on the structure of the term A_Y and uses Lemma 6.2.1.   ∎

   This completes the informal overview of the proof of the Church–Rosser theorem. To summarize, the Church–Rosser theorem was stated and a proof of it was sketched by defining the notion of a walk and demonstrating the diamond property of walks. The above overview should serve as a guide to the formalization of this theorem in the Boyer–Moore logic which is described in the next section.

6.3   Formalizing the Lambda Calculus in Lisp

The lambda calculus discussed in Section 6.2 will be formally described in the Boyer–Moore logic in this section. This formalization along with the formal statement of the Church–Rosser theorem will be discussed here by means of the actual input to the Boyer–Moore theorem prover. The approach used to define the syntax and operations of the lambda calculus within the Boyer–Moore logic is quite general and can be applied to wide variety of formal systems. In the next section, a similar formalization will be carried out in a notation for terms that is different from the one used in Section 6.2. The unqualified use of the words 'constant', 'variable', and 'term' refers to the lambda calculus constants, variables and terms and not to those in the Boyer–Moore logic.

   Initially the Boyer–Moore theory contains axioms for literal atoms, natural numbers and lists. Constants in the lambda calculus are represented by literal atoms and variables by natural numbers. Two new shells are added to represent lambda-abstractions and applications. Many of the functions described below are prefixed with the letter 'N' (for normal) to distinguish them from the notational variant defined in the next section. (Note: This is not to be confused with the fact that in Lisp, the prefix 'N' usually means 'NOT'.)

The shell used to represent lambda-abstractions consists of a 2-place constructor NLAMBDA whose first argument is coerced into being a lambda calculus variable.[4] If X represents x and Y represents Y, then (NLAMBDA X Y) would represent $(\lambda x.\ Y)$. The two corresponding destructor functions are NBIND and NBODY. Thus (NBIND (NLAMBDA X Y)) returns X if X is a number, and the default value 0 otherwise. (NBODY (NLAMBDA X Y)) returns Y. The recognizer for the shell is the 1-place function NLAMBDAP, and (NLAMBDAP (NLAMBDA X Y)) returns T.

Applications are represented by another shell in which the constructor is the 2-place function NCOMB. If X represents X and Y represents Y, then (NCOMB X Y) would represent $(X\ Y)$. The two destructors are NLEFT and NRIGHT, respectively. As expected, (NLEFT (NCOMB X Y)) is X, and (NRIGHT (NCOMB X Y)) is Y. The recognizer for the shell is NCOMBP, where (NCOMBP (NCOMB X Y)) is T.

Examples of terms

1. $((\lambda x.\ (x\ x))\ (\lambda x.\ (x\ x)))$

```
(NCOMB (NLAMBDA 1 (NCOMB 1 1))
       (NLAMBDA 1 (NCOMB 1 1)))
```

2. $(\lambda x.\ (a\ x))$

```
(NLAMBDA 0 (NCOMB 'A 0))
```

3. $(\lambda x.\ (\lambda y.\ (\lambda z.\ ((x\ z)\ (y\ z)))))$

```
(NLAMBDA 1
    (NLAMBDA 2 (NLAMBDA 3
             (NCOMB (NCOMB 1 3)
                    (NCOMB 2 3)))))
```

Having described the manner in which terms are represented, we can now examine the function that checks whether a given object is indeed a term. The function NTERMP displayed below carries out this task and returns T or F depending on whether the argument X is a term or not. When (NLAMBDAP X) holds, then X is of the form (NLAMBDA Y Z) and NTERMP recursively checks if the body (NBODY X) is a term. Note that since (NBIND X) always returns a variable, this does not need to be explicitly checked. Similarly, when X is an NCOMBP (i.e., of the form (NCOMB Y Z)), NTERMP is recursively invoked on the left and right subterms, respectively. If the given X is neither an abstraction nor an application, then X must either be a variable, in which case (NUMBERP X) is T, or a literal atom and (LITATOM X) is T.

---

[4]By the shell principle (see page 29), a shell is specified by a constructor function, a recognizer function, and zero or more accessor functions. One may in addition restrict the type of the value returned by an accessor function. For example, the accessor SUB1 corresponding to the constructor ADD1 must return a value of type NUMBERP. Thus, (SUB1 (ADD1 NIL)) would have to be equal to a specified default bottom object which is (ZERO) in this case.

```
Definition.
  (NTERMP X)
    =
(IF (NLAMBDAP X)
    (NTERMP (NBODY X))
    (IF (NCOMBP X)
        (AND (NTERMP (NLEFT X))
             (NTERMP (NRIGHT X)))
        (OR (NUMBERP X) (LITATOM X))))
```

The next step is to define the operation of substitution. The function NSUBST takes three arguments X, Y and N and substitutes Y for all occurrences of the variable N in X. The definition is identical to the one given in Section 6.1.1. In the case when X is an abstraction, NSUBST is invoked on (NBODY X) with Y and N provided (NBIND X) is not N. If (NBIND X) is N, then NSUBST returns X itself. When X is an application, NSUBST is invoked on (NLEFT X) and (NRIGHT X) respectively, with Y and N. Finally, NSUBST returns Y if X is a variable that is identical to N, and X otherwise.

```
Definition.
  (NSUBST X Y N)
    =
(IF (NLAMBDAP X)
    (IF (EQUAL (NBIND X) N)
        X
        (NLAMBDA (NBIND X)
                 (NSUBST (NBODY X) Y N)))
    (IF (NCOMBP X)
        (NCOMB (NSUBST (NLEFT X) Y N)
               (NSUBST (NRIGHT X) Y N))
        (IF (NUMBERP X)
            (IF (EQUAL X N) Y X)
            X)))
```

Examples

1. $[a/x](\lambda x.\ x) \equiv (\lambda x.\ x)$

   ```
   (NSUBST (NLAMBDA 1 1) 'A 1)
   = (NLAMBDA 1 1)
   ```

2. $[(\lambda x.\ (x\ x))/x](x\ x) \equiv ((\lambda x.\ (x\ x))\ (\lambda x.\ (x\ x)))$

   ```
   (NSUBST  (NCOMB 1 1)
            (NLAMBDA 1 (NCOMB 1 1)) 1)
     =
   (NCOMB (NLAMBDA 1 (NCOMB 1 1))
          (NLAMBDA 1 (NCOMB 1 1)))
   ```

3. $[(\lambda x.\ y)/x](\lambda y.\ x) \equiv (\lambda y.\ (\lambda x.\ y))$

   ```
   (NSUBST (NLAMBDA 2 1) (NLAMBDA 1 2) 1)
   = (NLAMBDA 2 (NLAMBDA 1 2))
   ```

(Note: the free variable y is captured in this substitution.)

The function NOT-FREE-IN used as (NOT-FREE-IN X Y) checks if the variable X does not occur free in the term Y. This is used to define FREE-FOR such that (FREE-FOR X Y) checks whether the term Y is free for any of the variables in the term X. The definitions of these two functions will be omitted.

The next definition provides a formal definition of an α-step. The function ALPHA-EQUAL checks if two terms are identical except for bound variables being renamed. The function takes four arguments: the two terms A and B, and two lists of bound variables X and Y corresponding to A and B respectively. ALPHA-EQUAL simultaneously traverses the terms A and B and checks whether they both have the same structure. ALPHA-EQUAL also collects the bound variables encountered as (NBIND ...) in A into X, and the bound variables in B into Y. When the recursion reaches the base case and a variable is encountered in A and in B, ALPHA-EQUAL checks whether the least index (corresponding to the *nearest* NLAMBDA binding that variable) of the variable from A in the list X is the same as the least index of the variable from B in Y. A function INDEX is used to return the least index of a variable in a list. Since variables are represented by numbers, if the given variable does not occur in the given list, INDEX returns one plus the sum of the length of the list and the number representing the variable. If a variable in A does not occur in the corresponding list of bound variables Y, then it must occur free in the original A. Assuming that X and Y are of the same length, if the INDEX of a free variable on X is the same as the INDEX of a free variable on Y, then they must be the same free variable. A is said to go to B in an α-step if (ALPHA-EQUAL A B NIL NIL) returns T. Note that this definition of "congruence modulo bound variables" corresponds closely to our intuitive notion of this relation.

```
Definition.
  (INDEX N LIST)
       =
  (IF (LISTP LIST)
      (IF (EQUAL (CAR LIST) N)
          1
          (ADD1 (INDEX N (CDR LIST))))
      (ADD1 N))
```

```
Definition.
  (ALPHA-EQUAL A B X Y)
       =
  (IF (AND (NLAMBDAP A) (NLAMBDAP B))
      (ALPHA-EQUAL (NBODY A)
                   (NBODY B)
                   (CONS (NBIND A) X)
                   (CONS (NBIND B) Y))
      (IF (AND (NCOMBP A) (NCOMBP B))
          (AND (ALPHA-EQUAL (NLEFT A) (NLEFT B) X Y)
               (ALPHA-EQUAL (NRIGHT A) (NRIGHT B) X Y))
          (IF (AND (NUMBERP A) (NUMBERP B))
              (EQUAL (INDEX A X) (INDEX B Y))
              (EQUAL A B))))
```

Examples

1. $((\lambda x.\ x)\ z) \overset{\alpha}{\mapsto} ((\lambda y.\ y)\ z)$

```
(ALPHA-EQUAL (NCOMB (NLAMBDA 1 1) 3)
             (NCOMB (NLAMBDA 2 2) 3)  NIL NIL)
```

2. $(\lambda x.\ (x\ (\lambda y.\ (x\ y)))) \overset{\alpha}{\mapsto} (\lambda y.\ (y\ (\lambda x.\ (y\ x))))$

```
(ALPHA-EQUAL (NLAMBDA 1 (NCOMB 1 (NLAMBDA 2 (NCOMB 1 2))))
             (NLAMBDA 2 (NCOMB 2 (NLAMBDA 1 (NCOMB 2 1))))
             NIL NIL)
```

3. $((\lambda x.\ (\lambda y.\ (x\ y)))\ (\lambda z.\ (y\ z))) \overset{\alpha}{\mapsto} ((\lambda x.\ (\lambda z.\ (x\ z)))\ (\lambda z.\ (y\ z)))$

```
(ALPHA-EQUAL (NCOMB  (NLAMBDA 1 (NLAMBDA 2 (NCOMB 1 2)))
                     (NLAMBDA 3 (NCOMB 2 3)))
             (NCOMB  (NLAMBDA 1 (NLAMBDA 3 (NCOMB 1 3)))
                     (NLAMBDA 3 (NCOMB 2 3)))
             NIL NIL)
```

The next important definition captures the relation $A \overset{\beta}{\mapsto} B$. The function NBETA-STEP takes two arguments: the terms A and B. It returns T if A and B are identical. If both A and B are abstractions, then NBETA-STEP is recursively invoked on their bodies. On the other hand, if (NCOMBP A) is T, then A could be a redex (if (NLAMBDAP (NLEFT A)) is T). Further, if (FREE-FOR (NBODY (NLEFT A)) (NRIGHT A)) is T and B is the result of substituting (NRIGHT A) for (NBIND (NLEFT A)) in (NBODY (NLEFT A)), then NBETA-STEP returns T. If A is not a redex, then (NCOMBP B) must be T, and NBETA-STEP is recursively invoked once on the left subterms of A and B, and once on the right subterms of A and B. If both recursive calls return T, then NBETA-STEP returns T. If none of the above cases is applicable, then NBETA-STEP returns F, since if the computation reached this point, A and B could not have been identical.

```
Definition.
  (NBETA-STEP A B)
    =
  (IF (EQUAL A B)
      T
      (IF (NLAMBDAP A)
          (AND (NLAMBDAP B)
               (EQUAL (NBIND A) (NBIND B))
               (NBETA-STEP (NBODY A) (NBODY B)))
          (IF (NCOMBP A)
              (OR (AND (NLAMBDAP (NLEFT A))
                       (FREE-FOR (NBODY (NLEFT A))
                                 (NRIGHT A))
                       (EQUAL B
                              (NSUBST (NBODY (NLEFT A))
                                      (NRIGHT A)
                                      (NBIND (NLEFT A)))))
                  (AND (NCOMBP B)
                       (NBETA-STEP (NLEFT A) (NLEFT B))
                       (NBETA-STEP (NRIGHT A) (NRIGHT B))))
              F)))
```

Examples

1. $((\lambda x.\,(a\;x))\;b) \overset{\beta}{\mapsto} (a\;b)$

   ```
   (NBETA-STEP (NCOMB (NLAMBDA 1 (NCOMB 'A 1)) 'B)
               (NCOMB 'A 'B))
   ```

2. $((\lambda y.\,((\lambda x.\,x)\;y))\;((\lambda x.\,(x\;x))\;a)) \overset{\beta}{\mapsto} ((\lambda x.\,x)\;((\lambda x.\,(x\;x))\;a))$

   ```
   (NBETA-STEP (NCOMB (NLAMBDA 2 (NCOMB (NLAMBDA 1 1) 2))
                      (NCOMB (NLAMBDA 1 (NCOMB 1 1)) 'A))
               (NCOMB (NLAMBDA 1 1)(NCOMB (NLAMBDA 1 (NCOMB 1 1)) 'A)))
   ```

3. $((\lambda y.\,((\lambda x.\,x)\;(b\;y)))\;((\lambda x.\,(x\;x))\;a)) \overset{\beta}{\mapsto} ((\lambda y.\,(b\;y))\;(a\;a))$

   ```
   (NBETA-STEP (NCOMB (NLAMBDA 2 (NCOMB (NLAMBDA 1 1) (NCOMB 'B 2)))
                      (NCOMB (NLAMBDA 1 (NCOMB 1 1)) 'A))
               (NCOMB (NLAMBDA 2 (NCOMB 'B 2)) (NCOMB 'A 'A)))
   ```

The next definition is that of a reduction and is quite straightforward. The function NREDUCTION checks if A reduces to B *via* the terms in LIST by means of α- or β-steps. The function NSTEP checks whether A goes to B in a single α- or β-step. In the definition of NREDUCTION, if LIST is empty, then (NSTEP A B) must be T for NREDUCTION to return T. Otherwise, NREDUCTION checks whether the first element of LIST (i.e., (CAR LIST)) goes to B in a β-step, and A reduces to the (CAR LIST) *via* (CDR LIST). Thus LIST represents the trace of the reduction from A to B in reverse order.

```
Definition.
  (NSTEP A B)

  =

  (OR (ALPHA-EQUAL A B NIL NIL)
      (NBETA-STEP A B))
```

```
Definition.
  (NREDUCTION A B LIST)

  =

  (IF (LISTP LIST)
      (AND (NSTEP (CAR LIST) B)
           (NREDUCTION A (CAR LIST) (CDR LIST)))
      (NSTEP A B))
```

Finally, the statement of the Church–Rosser theorem expressed in terms of this formalization would read:

```
∃REDS-Y, REDS-Z, W:
  (IMPLIES (AND (NTERMP X)
                (NREDUCTION X Y LIST1)
                (NREDUCTION X Z LIST2))
           (AND (NREDUCTION Y W REDS-Y)
                (NREDUCTION Z W REDS-Z)))
```

In English, if X reduces to Y *via* LIST1, and to Z *via* LIST2, then there exist REDS-Y, REDS-Z, and W such that Y reduces to W *via* REDS-Y and Z also reduces to W *via* REDS-Z. Since the Boyer–Moore logic does not permit the use of quantifiers, REDS-Y, REDS-Z, and W have to be explicitly constructed for any given X, Y, Z, LIST1, and LIST2. The same function can be used to construct both REDS-Y and REDS-Z since Y and Z, and LIST1 and LIST2, can be interchanged without changing the content of the above statement. The term (MAKE-N-W X Z Y LIST2 LIST1) constructs the required W. The term (NMAKE-REDUCTION X Y Z LIST1 LIST2) constructs REDS-Y, and (NMAKE-REDUCTION X Z Y LIST2 LIST1) constructs REDS-Z. The definitions of MAKE-N-W and NMAKE-REDUCTION are omitted. It must be noted that the actual definitions of these *witness* functions are not needed to check if the Church–Rosser theorem has been correctly captured in the statement below. We only need to know that these functions have been defined and their definitions have been accepted by the theorem prover. The statement of the Church–Rosser theorem as communicated to the theorem prover is as shown below.

```
Theorem.  FINALLY-CHURCH-ROSSER (rewrite):
  (IMPLIES (AND (NTERMP X)
                (NREDUCTION X Y LIST1)
                (NREDUCTION X Z LIST2))
           (AND (NREDUCTION Y
                           (MAKE-N-W X Z Y LIST2 LIST1)
                           (NMAKE-REDUCTION X Y Z
                                            LIST1 LIST2))
                (NREDUCTION Z
                           (MAKE-N-W X Z Y LIST2 LIST1)
                           (NMAKE-REDUCTION X Z Y
                                            LIST2 LIST1)))))
```

Thus, the lambda calculus as described in Section 6.1 has been formalized in the Boyer–Moore logic. The statement of the Church–Rosser theorem for this formalization has also been displayed. The proof of this statement has not been discussed. The proof was actually carried out by mapping this formalization into a formalization which uses a different notation for lambda calculus and then carrying out the proof in this new formalization. This formalization will be discussed in the next section.

6.4   Using the de Bruijn Notation

In this section, the lambda calculus will be formally defined using a notation due to de Bruijn, and the Church–Rosser theorem for this notation will be formally stated. The de Bruijn notation is interesting in that it obviates the need for α-steps, so that the proof turns out to be quite similar to the one sketched in Section 6.2. The formal description consists of the following parts:

1. The de Bruijn notation for lambda-terms.

2. Definition of beta-reduction.

3. Definition of a walk.

4. Statement of the diamond property for walks.

5. Statement of the Church–Rosser theorem.

6.4.1   The de Bruijn Notation

As mentioned earlier, a part of the mechanical proof employs a notation for lambda calculus terms that is different from the standard notation introduced in Section 6.1 and formalized in Section 6.3. The reasons for using this notation are discussed in Section 6.7. The notation is due to de Bruijn [dB72] and is employed for proof-checking purposes in the Au-TOMATH family of proof-checkers.[5] A description of this notation appears in Appendix C of Barendregt's *The Lambda Calculus* [Bar78a]. The next few paragraphs explain this notation briefly and attempt to convince the reader that no essential part of lambda calculus is lost by switching to this notation. The equivalence of the de Bruijn and the standard notations been proved as part of the mechanical proof.

The de Bruijn notation is significantly different from the standard notation used in Sections 6.1, 6.2 and 6.3. A fair amount of confusion can arise from trying to carry over intuitions based on one notation to the other. In the standard notation, distinct variables have distinct names and an occurrence of a variable can be identified by the appearance of its name. We can also distinguish between free and bound occurrences of a variable in a given term. More strongly, given an occurrence of a variable, we can recognize if it is bound and identify the λ that binds it. If an occurrence of the variable x is free in the subterm M, then it is bound by the underlined λ in the term $(\underline{\lambda}x.\ M)$. Correspondingly, an occurrence of x is bound by the closest *surrounding* λ that is *tagged* with the name x, where the λ in λx is said to be tagged by x, and the underlined λ in $(\underline{\lambda}x.\ M)$ surrounds every variable occurrence in M.

The de Bruijn notation used in this proof does away with names for variables and tags for λ, while retaining the ability to determine whether and where a particular variable occurrence is bound. The idea is quite simple. A variable occurrence is represented by a positive integer, say n, indicating that the variable occurrence is bound by the n-th surrounding λ. For example, the underlined λ in the term $(\underline{\lambda}\ (\underline{1}\ (\lambda\ (1\ \underline{2}))))$ binds the underlined variable occurrences because it is the first surrounding λ from the underlined 1, but the second surrounding λ from the underlined 2. Applications of the form $(X\ Y)$ have the same syntax as in the standard notation. If corresponding to a variable occurrence represented by n in a subterm M, there are fewer than n surrounding λs, then that variable occurrence is free in M. For example, the underlined variable occurrences are free in $((\lambda\ (\underline{2}\ (\lambda\ (2\ \underline{3}))))\ (\lambda\ (\underline{4}\ 1)))$.

One important difference between the standard notation and the de Bruijn notation is that in the de Bruijn notation, if M is a subterm of X, then the order in which the free variable occurrences in M can become bound in X is fixed regardless of X. Taking the previous example, if $((\lambda\ (\underline{2}\ (\lambda\ (2\ \underline{3}))))\ (\lambda\ (\underline{4}\ 1)))$ is a subterm of a term in which the underlined variables 2, 3 and 4 become bound, then the λ binding 2 and 3 must be contained within or surrounded by the λ that binds 4. For this reason, it is helpful to always think of terms as being subterms in some bigger term in which all the free variables become bound.

Another important difference is that since variables do not have names in the de Bruijn

---

[5]The paper [dB72] introducing the de Bruijn notation also describes an elegant proof of the Church–Rosser theorem which is similar to the one described here. I thank Professor de Bruijn for drawing my attention to it.

notation, it makes no sense to rename bound variables as was done in the standard notation using an α-step.

The representation of these terms in the Boyer–Moore logic is quite similar to that described in the previous section. Two new shells are added to represent terms of the form $(\lambda\ M)$ and $(M\ N)$ respectively. The first is a shell with a 1-place recognizer LAMBDAP, a 1-place constructor LAMBDA, and a 1-place destructor BODY. (LAMBDAP (LAMBDA X)) is T, and (BODY (LAMBDA X)) is X.

The second shell consists of a 1-place recognizer COMBP, a 2-place constructor COMB, and two 1-place destructors: LEFT and RIGHT. (COMBP (COMB X Y)) is T, (LEFT (COMB X Y)) returns X, and (RIGHT (COMB X Y)) returns Y.

Variables are represented by the positive natural numbers 1, 2, 3, etc. The literal atoms in the Boyer–Moore logic are used to represent constants. A few examples of lambda-terms represented as objects in the Boyer–Moore logic are given below along with their counterparts in the standard notation.

Examples

1. $(\lambda x.\ x)$
   (LAMBDA 1)

2. $(\lambda x.\ (\lambda y.\ (\lambda z.\ (x\ (y\ z)))))$
   (LAMBDA (LAMBDA (LAMBDA (COMB 3 (COMB 2 1)))))

3. $(\lambda x.\ ((\lambda y.\ (x\ (y\ z)))\ (x\ z)))$
   (LAMBDA (COMB (LAMBDA (COMB 2 (COMB 1 3))) (COMB 1 2)))

We can now examine the definition of the function which translates a term in the standard notation to the corresponding term in the de Bruijn notation. The function TRANSLATE below bears an obvious resemblance to the function ALPHA–EQUAL defined in the previous section. TRANSLATE takes two arguments: the term to be translated, X, and a list of bound variables, BOUNDS. It works by recursion on the structure of the term X and collects the bound variables thus encountered into BOUNDS. When, in the base case, a variable occurrence is encountered, TRANSLATE returns the INDEX of X in BOUNDS. The function INDEX has been described in the previous section. Any two alpha-equal terms have the same translation, and two terms which have the same translation must be alpha-equal.

```
Definition.
  (TRANSLATE X BOUNDS)
  =
  (IF (NLAMBDAP X)
      (LAMBDA (TRANSLATE (NBODY X)
                         (CONS (NBIND X) BOUNDS)))
      (IF (NCOMBP X)
          (COMB (TRANSLATE (NLEFT X) BOUNDS)
                (TRANSLATE (NRIGHT X) BOUNDS))
          (IF (NUMBERP X) (INDEX X BOUNDS) X)))
```

6.4.2   Beta-Reduction

The definition of beta-reduction in this notation is substantially more complicated than the corresponding definition in the standard notation. This is mainly because it ensures that no free variables get captured during substitution. This definition turned out to be the crucial step in the proof.

As before, a subterm of the form (COMB (LAMBDA X) Y) is labelled a *redex*. A beta-reduction should transform this subterm into the subterm got by replacing all those occurrences in X of the variable bound by the outermost LAMBDA in (LAMBDA X) with an *appropriate transformation* of Y. It will turn out that the free variables in (LAMBDA X) as well as those in Y can be affected during beta-reduction. The reason we need an appropriate transformation of Y is to ensure that the free variables in Y do not become bound in the result of the beta-reduction.

As an example, consider a redex of the form

$$\text{(COMB (LAMBDA (LAMBDA 2)) 1)}.$$

By the definition of binding, 2 is bound by the outermost LAMBDA in (LAMBDA (LAMBDA 2)). So, 1 must be substituted for 2 in (LAMBDA 2). If this is done in the naive way, the result is (LAMBDA 1). The variable 1 which was free in the original term becomes bound in the result of the beta-reduction. The correct outcome of this beta-reduction should be (LAMBDA 2) so that the variable 1 in the original term is transformed to 2 in the result. Then, if this redex is part of a larger term, say

$$\underline{\text{(LAMBDA}} \text{ (COMB (LAMBDA (LAMBDA 2)) 1)),}$$

then the transformed version of 1, namely, 2, remains bound to the underlined LAMBDA in the result of the beta-reduction, (<u>LAMBDA</u> (LAMBDA 2)). More generally, while substituting Y into the body of a term (LAMBDA M), all the free variables in Y must be incremented by one. The definition of the function that carries out the operation of incrementing all the free variables is discussed below.

The function BUMP takes two arguments, a term X and a counter N. BUMP goes through the term and returns the same term but with the free variables of level greater than N, incremented by one. N should be 0 in the initial call to ensure that all free variables get incremented.

```
Definition.
  (BUMP X N)
      =
  (IF (LAMBDAP X)
      (LAMBDA (BUMP (BODY X) (ADD1 N)))
      (IF (COMBP X)
          (COMB (BUMP (LEFT X) N)
                (BUMP (RIGHT X) N))
          (IF (LESSP N X) (ADD1 X) X)))
```

Examples of BUMP

1. (BUMP (LAMBDA (LAMBDA (COMB 2 3))) 0)
   = (LAMBDA (LAMBDA (COMB 2 4)))

2. (BUMP (LAMBDA (LAMBDA (COMB 1 2))) 0)
   = (LAMBDA (LAMBDA (COMB 1 2)))

3. (BUMP (LAMBDA (COMB (LAMBDA 3)(COMB 3 1))) 1)
   = (LAMBDA (COMB (LAMBDA 3)(COMB 4 1)))

Now we can use the function BUMP to define beta-reduction by means of a function SUBST which is similar but not identical to the substitution operation defined as NSUBST (page 6.3). There are several important differences and SUBST should not be mistaken for substitution as previously defined. It was mentioned earlier that when reducing the redex (COMB (LAMBDA X) Y), the free variables of both (LAMBDA X) and Y are affected. The free variables of Y are affected because they might come to be surrounded by additional LAMBDAs. This is remedied by incrementing all the free variables in Y (using BUMP) every time SUBST is recursively invoked on the body of a lambda-abstraction.

The free variables in the (LAMBDA X) part of a redex (COMB (LAMBDA X) Y) are affected because the LAMBDA in (LAMBDA X) is lost in the course of beta-reduction. This means that this LAMBDA no longer surrounds the free variables in (LAMBDA X). For example, consider the beta-reduction of the redex

$$\text{(COMB (LAMBDA (COMB 1 2)) 1)}$$

occurring in the term

$$\text{(\underline{LAMBDA} (COMB (LAMBDA (COMB 1 \underline{2})) \underline{1}))}$$

in which the underlined variables are bound by the underlined LAMBDA. If this is carried out in the naive fashion, the result is

$$\text{(\underline{LAMBDA} (COMB \underline{1} \underline{2}))}$$

in which the underlined symbols are the same as the underlined symbols in the original term. However, in this result, only the underlined 1 is bound by the underlined LAMBDA. This means the underlined variable 2 lost its original binding due to the fact that a LAMBDA was eliminated by the beta-reduction. If done correctly, the 2 should have been decremented by one in the result yielding (LAMBDA (COMB 1 1)). More generally, all the free variables in (LAMBDA X) will have to be decremented by one when beta-reducing a redex of the form (COMB (LAMBDA X) Y). The definition of SUBST below takes the above factors into consideration. It is important to note that the name SUBST for this function is not a misnomer. Since a free variable in X is being replaced, all the free variables in X of higher level are decremented by one.

The function SUBST takes three arguments: two terms X and Y (corresponding to the X and Y in the redex (COMB (LAMBDA X) Y)), and a counter N. The counter is used in two ways. The first is to indicate the occurrences of the variable bound by the LAMBDA in (LAMBDA X) so that they can be replaced by the appropriate transformation of Y. The second way is to indicate the free variables in (LAMBDA X) so that they can be decremented for the reason given in the previous paragraph.

```
Definition.
  (SUBST X Y N)
    =
  (IF (LAMBDAP X)
      (LAMBDA (SUBST (BODY X) (BUMP Y 0) (ADD1 N)))
      (IF (COMBP X)
          (COMB (SUBST (LEFT X) Y N)
                (SUBST (RIGHT X) Y N))
          (IF (NOT (ZEROP X))
              (IF (EQUAL X N)
                  Y
                  (IF (LESSP N X) (SUB1 X) X))
              X)))
```

In the base case, X is either a constant or a variable. The test (NOT (ZEROP X)) is T only if X is a positive natural number (i.e., a variable). If (EQUAL X N) is T, then X is the variable bound by the LAMBDA in the redex and it is replaced by Y. If (EQUAL X N) is F, then X is either a free variable or a bound one. If (LESSP N X) is T, then X is a free variable in the original redex and it is decremented by one. Otherwise, X is a bound variable and it is returned unchanged. On the other hand, if (NOT (ZEROP X)) is F, then X is a constant and it is returned unchanged.

If X is of the form (LAMBDA M), then SUBST returns (LAMBDA M1), where M1 is the result of the recursive call of SUBST on the M. In the recursive call (BUMP Y 0) replaces Y and (ADD1 N) replaces N. The reason for incrementing all the free variables in Y by one when substituting into (BODY X) was explained earlier. The reason for replacing N with (ADD1 N) is to record the additional surrounding LAMBDA.

The case when X is of the form (COMB U V) turns out to be quite simple. SUBST is invoked on U and V, returning U1 and V1, respectively. The parameters Y and N remain unchanged in both recursive calls. (COMB U1 V1) is returned as the result.

Examples of SUBST

1.   (SUBST 1 (LAMBDA 1) 1) = (LAMBDA 1)

2.   (SUBST (LAMBDA (COMB 2 3)) (LAMBDA (COMB 1 2)) 1)

   =

   (LAMBDA (COMB (LAMBDA (COMB 1 3)) 2))

3.   (SUBST (COMB (LAMBDA (LAMBDA (COMB 1 (COMB 4 5))))
                    (COMB (LAMBDA 3)(LAMBDA 4)))
            1 2)

   =

   (COMB (LAMBDA (LAMBDA (COMB 1 (COMB 3 4))))
         (COMB (LAMBDA 2)(LAMBDA 3)))

(SUBST X Y 1) denotes the result of beta-reducing the redex (COMB (LAMBDA X) Y). The theorem TRANSLATE-PRESERVES-REDUCTION below states that SUBST and NSUBST correspond in the context of a beta-reduction.

```
Theorem.  TRANSLATE-PRESERVES-REDUCTION (rewrite):
   (IMPLIES (AND (NLAMBDAP X)
                 (FREE-FOR (NBODY X) Y)
                 (NTERMP X)
                 (NTERMP Y))
            (EQUAL (TRANSLATE (NSUBST (NBODY X) Y (NBIND X))
                              BOUNDS)
                   (SUBST (BODY (TRANSLATE X BOUNDS))
                          (TRANSLATE Y BOUNDS)
                          1)))
```

This completes the description of beta-reduction. We can now use this to describe the notion of a walk.

6.4.3 Definition of a Walk

In the informal proof-sketch, a walk was described as a sequence of beta-reductions on the redexes of a term X, which were constrained so that the reductions on overlapping redexes took place in an inside-out order. A relation $X \mapsto Y$ was defined to capture the notion of a walk. In the formalization below, this relation will be replaced by a function WALK which performs the beta-reductions on X in the appropriate order to yield Y. The reductions to be performed are indicated by an object called a *walk-instruction*. If WALK applies a walk-instruction to a term X and transforms it to Y, then $X \mapsto Y$ will hold. Conversely, for any pair of terms X, Y such that $X \mapsto Y$, there will exist a walk-instruction that takes X to Y. Intuitively, if X is the term to be beta-reduced, a representation of X in which the redexes to be reduced are underlined could serve as a walk-instruction. Then, to walk X, one would only have to beta-reduce the underlined redexes in the walk-instruction while ensuring that all the underlined subterms within it have already been reduced. A much simpler form of walk-instruction is actually used. The walk-instruction used below is a list-structure that has approximately the same structure as the term to be beta-reduced. That is, a walk-instruction is a triple consisting of a left and a right walk-instruction, and a *command*. If the command 'REDUCE (the Lisp atom "REDUCE") is encountered in the walk-instruction where a redex $((\lambda x.\ M)\ N)$ appears in the term, WALK returns $[N'/x]M'$, where $(\lambda x.\ M')$ is the result of applying the left walk-instruction to $(\lambda x.\ M)$, and N' is the result of applying the right walk-instruction to N. So, the command 'REDUCE plays the role of the underlining below a redex.

The function WALK below applies a walk-instruction W to a term M to return the conclusion of the walk. There are no syntactic restrictions on what constitutes a valid walk-instruction. If M is of the form (LAMBDA X), then WALK applies W to X. If M is a redex of the form (COMB (LAMBDA X) Y), and W is a walk-instruction of the form ⟨'REDUCE W-LEFT W-RIGHT⟩, then WALK beta-reduces the redex obtained by recursively applying W-LEFT to X and W-RIGHT to Y. If M is of the form (COMB X Y) but is either not a redex or the command field of W is not 'REDUCE, then WALK returns (COMB X1 Y1), where X1 and Y1 are the results of W-LEFT applied to X1 and W-RIGHT applied to Y1, respectively. In the base case, when M is an atom, M is returned unchanged. The definition below may be compared with the definition of the relation $X \mapsto Y$ in the informal outline.

```
Definition.
  (WALK W M)
    =
  (IF (LAMBDAP M)
      (LAMBDA (WALK W (BODY M)))
      (IF (COMBP M)
          (IF (AND (EQUAL (COMMAND W) 'REDUCE)
                   (LAMBDAP (LEFT M)))
              (SUBST (BODY (WALK (LEFT-INSTRS W) (LEFT M)))
                     (WALK (RIGHT-INSTRS W) (RIGHT M))
                     1)
              (COMB (WALK (LEFT-INSTRS W) (LEFT M))
                    (WALK (RIGHT-INSTRS W) (RIGHT M))))
          M))
```

The above notion of a walk can now be used to formally state the diamond property for walks in the Boyer–Moore logic.

6.4.4   The Statement of the Diamond Property for Walks

The statement of the diamond property from Section 6.2 was as follows:

If $X \mapsto Y$ and $X \mapsto Z$, then there exists a W such that $Y \mapsto W$ and $Z \mapsto W$.

The task here is to restate the diamond property using the function WALK instead of the relation $X \mapsto Y$. Let U and V be the two walk-instructions which when applied to the term X lead to the terms Y and Z respectively, in the statement above. Now, instead of asserting the existence of a term W to which both Y and Z can be made to *converge* by means of walks, we can assert the existence of two *convergent* walk-instructions. In other words, there exist walk-instructions W1 and W2, such that the result of applying W1 to the result of applying U to X, is the same as the result of applying W2 to the result of applying V to X. Denoting X by M, this translates to:

```
∃W1, W2 (EQUAL (WALK W1 (WALK U M))
               (WALK W2 (WALK V M))).
```

Since the Boyer–Moore logic does not permit the use of quantifiers, the existential quantifiers over W1 and W2 will have to be replaced by functions of M, U and V. Let W1 be replaced by the term (MAKE-WALK M U V). Then, by symmetry, W2 can be replaced by (MAKE-WALK M V U). These replacements render the statement of the diamond property for walks into the following form:

```
Theorem.  MAIN (rewrite):
  (EQUAL (WALK (MAKE-WALK M U V) (WALK U M))
         (WALK (MAKE-WALK M V U) (WALK V M)))
```

Note that the actual definition of MAKE-WALK is not relevant to the argument that the above lemma correctly expresses the diamond property for walks. The statement of the Church–Rosser theorem will be similar in form to the statement of the diamond property for walks.

6.4.5   The Statement of the Church–Rosser Theorem

At this point, two assumptions will be made. The first is that any single β-step can be represented as a walk. The second assumption is that any walk can be represented as a sequence of beta-reductions. These are fairly self-evident observations, and they have been mechanically verified. A sequence of walks starting from a term M will be defined by repeatedly applying the walk-instructions from a given *list* of walk-instructions W using the function REDUCE. The definition of REDUCE given below is quite straightforward.

```
Definition.
  (REDUCE W M)
    =
  (IF (LISTP W)
      (REDUCE (CDR W) (WALK (CAR W) M))
      M)
```

Given the assumptions made in the previous paragraph, the statement of the Church–Rosser theorem can be expressed as:

$$\exists \; \text{W1, W2: (EQUAL (REDUCE W1 (REDUCE V M))}$$
$$\text{(REDUCE W2 (REDUCE U M)))}.$$

To relate the above statement to the one given in Section 6.2, M corresponds to X, (REDUCE U M) represents Y and (REDUCE V M) represents Z. W1 and W2 are the sequences of walk-instructions that close the diamond. The theorem asserts that there exist two sequences of walks represented by the walk-instruction sequences W1 and W2, such that W1 applied to Y is the same as W2 applied to Z, namely, the W in the statement of the theorem from Section 6.2. As before, the existential quantifiers over W1 and W2 will have to be replaced by functions. The term replacing W1 is (MAKE-REDUCE M U V). For reasons of symmetry, the term replacing W2 will be (MAKE-REDUCE M V U). After these replacements, the final statement of the Church–Rosser theorem is as below.

```
Theorem.  CHURCH-ROSSER (rewrite):
  (EQUAL (REDUCE (MAKE-REDUCE M U V)
                 (REDUCE V M))
         (REDUCE (MAKE-REDUCE M V U)
                 (REDUCE U M)))
```

This concludes the presentation of the formalization of the lambda calculus using the de Bruijn notation, within the Boyer–Moore logic. The de Bruijn notation for the lambda calculus was described and the beta-reduction operation was defined for it. Beta-reduction was used in defining the notion of a walk, and the diamond property for walks was formally stated. The definition of a walk was used in defining the relation X *reduces to* Y, which was in turn used to state the Church–Rosser theorem. The next step will be to discuss the proof of the Church–Rosser theorem in the form stated above. The definitions of MAKE-REDUCE and MAKE-WALK are only needed to construct the appropriate reductions and walks. The justification of the claim that the above statement of the Church–Rosser theorem is an acceptable one, depends only on the definitions leading up to the definition of REDUCE, and the two assumptions stated earlier.

6.5 Highlights of the Mechanical Proof

Having formally defined the lambda calculus and stated a version of the Church–Rosser theorem for it, we can now look at some of the theorems used in the mechanical proof. No proofs will be given in the description below. The theorems will be discussed in the following order:

1. Some lemmas on SUBST, WALK, and BUMP.

2. The substitutivity of walks lemma.

3. The diamond property for walks.

4. The Church–Rosser theorem.

6.5.1 Some Properties of BUMP, SUBST, and WALK

The lemmas discussed below state some interesting properties of the functions BUMP, SUBST, and WALK. The lemmas as stated are nct all obvious. This is because of the nature of the definition of SUBST.

The first of these is a straightforward lemma about commuting the order of two successive applications of BUMP to a term Y. In the statement below, the counter N has a value no greater than the value of counter M. The left-hand side of the conclusion has (ADD1 M) instead of M. The reason for that will become clear if the lemma is tested with N less than M and Y identical to M.

```
Theorem.  BUMP-BUMP (rewrite):
  (IMPLIES (LEQ N M)
           (EQUAL (BUMP (BUMP Y N) (ADD1 M))
                  (BUMP (BUMP Y M) N)))
```

The next couple of lemmas will relate the operations BUMP and SUBST. Both demonstrate how BUMP distributes over SUBST. The first one, BUMP-SUBST, deals with the case when the free variable being replaced by SUBST is first incremented due to BUMP. The reason for N being replaced by (ADD1 N) in the right-hand side of the conclusion will become clear if the lemma is tested with X identical to N. It is also interesting to test the lemma with X set to a number larger than N.

```
Theorem.  BUMP-SUBST (rewrite):
  (IMPLIES (LESSP M N)
           (EQUAL (BUMP (SUBST X Y N) M)
                  (SUBST (BUMP X M)
                         (BUMP Y M)
                         (ADD1 N))))
```

The other lemma relating BUMP and SUBST, ANOTHER-BUMP-SUBST deals with the case when the variable being substituted into does not get incremented by BUMP. A good example to test this lemma is when X is (ADD1 (ADD1 M)).

```
  Theorem.  ANOTHER-BUMP-SUBST (rewrite):
(IMPLIES (LEQ N (ADD1 M))
(EQUAL (SUBST (BUMP X (ADD1 M)) (BUMP Y M) N)
(BUMP (SUBST X Y N) M)))
```

There is another lemma about BUMP and SUBST which is displayed below as SUBST-NOT-FREE-IN. A good test for this lemma is when X is (ADD1 N).

```
Theorem.  SUBST-NOT-FREE-IN (rewrite):
    (EQUAL (SUBST (BUMP X N) Y (ADD1 N))
          X)
```

The next lemma is perhaps the most significant one in the proof. Just as BUMP-BUMP stated that the order of two successive BUMP operations could be exchanged, the lemma SUBST-SUBST below demonstrates how the order of two successive SUBST operations can be exchanged. The right-hand side of the conclusion shows the two successive SUBST operations. In the first, Y replaces the variable of level (ADD1 M) in X. In the second, the variable of level N in the result of the first SUBST is replaced by Z. Since M is less than N, the first SUBST would have decremented (ADD1 N) by one. Therefore, Z is actually replacing (ADD1 N) in X. The left-hand side reflects this. This lemma may be tested with (COMB 1 2) for X, 0 for M, and 1 for N, Y, and Z. The (NUMBERP M) antecedent could have been eliminated.

```
  Theorem.  SUBST-SUBST (rewrite):
    (IMPLIES (AND (NUMBERP M) (LESSP M N))
            (EQUAL (SUBST (SUBST X (BUMP Z M) (ADD1 N))
                          (SUBST Y Z N)
                          (ADD1 M))
                  (SUBST (SUBST X Y (ADD1 M)) Z N)))
```

At first sight, the statement of the above lemma might seem contrived and motivated by some specific requirement in the proof. The statement was however formulated prior to any detailed speculation regarding the rest of the proof. Testing the lemma on a few examples will clarify why this is a natural statement of the required property of SUBST.

6.5.2   The Substitutivity of Walks

The informal statement of the substitutivity of walks was given in Lemma 6.2.2 as:

If M walks to M', and N walks to N', then $[N/x]M$ walks to $[N'/x]M'$.

The same lemma stated below will read differently because it is stated in terms of walk-instructions rather than the *walks-to* relation. In English, it asserts:

If W and U are two walk-instructions such that W applied to M yields M', and U applied to X yields X', then there exists a walk-instruction V such that V applied to (SUBST M X N) yields (SUBST M' X' N).

In the statement below, the required walk-instruction V is constructed by the function SUB-WALK, and M' and X' are replaced by (WALK W M) and (WALK U X) respectively

```
Theorem.  WALK-SUBST (rewrite):
   (IMPLIES (NOT (ZEROP N))
            (EQUAL (WALK (SUB-WALK W M U N)
                        (SUBST M X N))
                   (SUBST (WALK W M) (WALK U X) N))))
```

As was seen in the proof-sketch in Section 6.2, the substitutivity of walks is the key to the proof of the diamond property of walks.

6.5.3   The Church–Rosser Theorem

Some assumptions were made in stating the Church–Rosser theorem as shown in page 166. These were that any β-step could be expressed as a walk, and that any walk could be expressed as a series of β-steps. These assumptions have been proved using the Boyer–Moore theorem prover in order to prove the Church–Rosser theorem for the de Bruijn notation in its original form. The formal statements of these assertions are presented below. The function BETA-STEP checks if its first argument goes to its second argument in a single β-step. The function REDUCTION checks if its first argument goes to the second argument through a list of β-steps *via* the terms listed in its third argument. The functions MAKE-WALK-STEP and MAKE-REDUCE-STEPS are used to construct the required walk-instruction and the list of terms in the reduction, respectively. The theorem MAKE-WALK-STEP-WALKS below asserts that if A goes to B in a β-step, then there exists a walk-instruction which walks A to B. The theorem MAKE-REDUCE-STEPS-STEPS shows that the relation between A and (REDUCE W A) can be expressed as a reduction.

```
Theorem.  MAKE-WALK-STEP-WALKS (rewrite):
   (IMPLIES (BETA-STEP A B)
            (EQUAL (WALK (MAKE-WALK-STEP A B) A)
                   B))
```

```
Theorem.  MAKE-REDUCE-STEPS-STEPS (rewrite):
   (REDUCTION A
              (REDUCE W A)
              (MAKE-REDUCE-STEPS W A))
```

Finally, the Church–Rosser theorem is proved in the acceptable form by the two lemmas below, which together demonstrate that if X goes to Y and Z through a series of β-steps, then there exists a W to which both Y and Z can be reduced. The function MAKE-W constructs the required W.

```
Theorem.  THE-REAL-CHURCH-ROSSER (rewrite):
   (IMPLIES (AND (REDUCTION X Y LIST1)
                 (REDUCTION X Z LIST2))
            (AND (REDUCTION Y
                           (MAKE-W X Z Y LIST2 LIST1)
                           (MAKE-REDUCTION X Y Z LIST1 LIST2))
                 (REDUCTION Z
                           (MAKE-W X Y Z LIST1 LIST2)
                           (MAKE-REDUCTION X Z Y LIST2 LIST1)))))
```

```
Theorem.  BOTH-MAKE-W-ARE-SAME (rewrite):
  (IMPLIES (AND (REDUCTION X Y LIST1)
                (REDUCTION X Z LIST2))
           (EQUAL (MAKE-W X Y Z LIST1 LIST2)
                  (MAKE-W X Z Y LIST2 LIST1)))
```

The theorems THE-REAL-CHURCH-ROSSER and BOTH-MAKE-W-ARE-SAME were used to prove the theorem FINALLY-CHURCH-ROSSER which was stated in Section 6.3, page 158. To do this, the X, Y, Z, LIST1, and LIST2 from FINALLY-CHURCH-ROSSER were translated into the de Bruijn notation using TRANSLATE, to yield the corresponding X, Y, Z, LIST1, and LIST2 in THE-REAL-CHURCH-ROSSER and BOTH-MAKE-W-ARE-SAME. The results of the functions MAKE-REDUCTION and MAKE-W were then translated back from the de Bruijn notation to the standard notation to get the corresponding NMAKE-REDUCTION and MAKE-N-W used in FINALLY-CHURCH-ROSSER.

The translation back into the standard notation was done using the function UNTRANS. The definition of UNTRANS is quite tricky. It takes three arguments: the term X, a number M that is at least the level of the highest free variable in X, and a counter N. The key idea behind the definition of UNTRANS is to maintain the separation between names for free variables and bound variables using M. Free variables get translated to one less than their level in X which is going to be a number less than M. The N-th λ is made to correspond to the bound variable M+N+1, which ensures its separation from the free variables in a fairly conservative way. By this definition of UNTRANS, the result of applying UNTRANS followed by TRANSLATE to a term leaves it unchanged.

```
Definition.
  (UNTRANS X M N)
    =
  (IF (LAMBDAP X)
      (NLAMBDA (PLUS M (ADD1 N))
               (UNTRANS (BODY X) M (ADD1 N)))
      (IF (COMBP X)
          (NCOMB (UNTRANS (LEFT X) M N)
                 (UNTRANS (RIGHT X) M N))
          (IF (LESSP N X)
              (DIFFERENCE X (ADD1 N))
              (IF (NOT (ZEROP X))
                  (ADD1 (PLUS M (DIFFERENCE N X)))
                  X))))
```

In summary, the lemmas leading to a proof of the Church–Rosser theorem were stated and proved using the Boyer–Moore theorem prover. This was achieved by first proving the theorem using the de Bruijn notation for lambda calculus terms and then mapping the result back into the standard notation. The proofs involved in the translation back and forth have been omitted from the discussion. The lemmas used in the proof included several interesting properties of the functions BUMP, SUBST, and WALK, the substitutivity of walks, and the diamond property of walks.

6.6 A Sample Mechanical Proof

The following is the output of the Boyer–Moore theorem prover when invoked on the lemma stating the diamond property of walks in the de Bruijn notation.

```
(PROVE-LEMMA MAIN
          (REWRITE)
          (EQUAL (WALK (MAKE-WALK M U V) (WALK U M))
                 (WALK (MAKE-WALK M V U) (WALK V M))))
```

Give the conjecture the name *1.

Perhaps we can prove it by induction. The recursive terms in the conjecture suggest four inductions. However, they merge into one likely candidate induction. We will induct according to the following scheme:

```
     (AND (IMPLIES (AND (LAMBDAP M) (p (BODY M) U V))
                   (p M U V))
          (IMPLIES (AND (NOT (LAMBDAP M))
                        (COMBP M)
                        (AND (EQUAL (COMMAND U) 'REDUCE)
                             (LAMBDAP (LEFT M)))
                        (p (LEFT M)
                           (LEFT-INSTRS U)
                           (LEFT-INSTRS V))
                        (p (RIGHT M)
                           (RIGHT-INSTRS U)
                           (RIGHT-INSTRS V)))
                   (p M U V))
          (IMPLIES (AND (NOT (LAMBDAP M))
                        (COMBP M)
                        (NOT (AND (EQUAL (COMMAND U) 'REDUCE)
                                  (LAMBDAP (LEFT M))))
                        (AND (EQUAL (COMMAND V) 'REDUCE)
                             (LAMBDAP (LEFT M)))
                        (p (LEFT M)
                           (LEFT-INSTRS U)
                           (LEFT-INSTRS V))
                        (p (RIGHT M)
                           (RIGHT-INSTRS U)
                           (RIGHT-INSTRS V)))
                   (p M U V))
          (IMPLIES (AND (NOT (LAMBDAP M))
                        (COMBP M)
                        (NOT (AND (EQUAL (COMMAND U) 'REDUCE)
                                  (LAMBDAP (LEFT M))))
                        (NOT (AND (EQUAL (COMMAND V) 'REDUCE)
                                  (LAMBDAP (LEFT M))))
                        (p (LEFT M)
                           (LEFT-INSTRS U)
                           (LEFT-INSTRS V))
                        (p (RIGHT M)
                           (RIGHT-INSTRS U)
                           (RIGHT-INSTRS V)))
```

```
                       (p M U V))
        (IMPLIES (AND (NOT (LAMBDAP M))
                      (NOT (COMBP M)))
                 (p M U V))).
```

Linear arithmetic and the lemmas BODY-LESSP, RIGHT-LESSP, and LEFT-LESSP establish that the measure (COUNT M) decreases according to the well-founded relation LESSP in each induction step of the scheme. Note, however, the inductive instances chosen for U and V. The above induction scheme leads to the following five new conjectures:

```
Case 5. (IMPLIES (AND (LAMBDAP M)
                      (EQUAL (WALK (MAKE-WALK (BODY M) U V)
                                   (WALK U (BODY M)))
                             (WALK (MAKE-WALK (BODY M) V U)
                                   (WALK V (BODY M)))))
                 (EQUAL (WALK (MAKE-WALK M U V) (WALK U M))
                        (WALK (MAKE-WALK M V U) (WALK V M)))).
```

This simplifies, appealing to the lemma BODY-LAMBDA, and unfolding MAKE-WALK and WALK, to:

```
        T.
```

```
Case 4. (IMPLIES (AND (NOT (LAMBDAP M))
                      (COMBP M)
                      (AND (EQUAL (COMMAND U) 'REDUCE)
                           (LAMBDAP (LEFT M)))
                      (EQUAL (WALK (MAKE-WALK (LEFT M)
                                              (LEFT-INSTRS U)
                                              (LEFT-INSTRS V))
                                   (WALK (LEFT-INSTRS U) (LEFT M)))
                             (WALK (MAKE-WALK (LEFT M)
                                              (LEFT-INSTRS V)
                                              (LEFT-INSTRS U))
                                   (WALK (LEFT-INSTRS V) (LEFT M))))
                      (EQUAL (WALK (MAKE-WALK (RIGHT M)
                                              (RIGHT-INSTRS U)
                                              (RIGHT-INSTRS V))
                                   (WALK (RIGHT-INSTRS U) (RIGHT M)))
                             (WALK (MAKE-WALK (RIGHT M)
                                              (RIGHT-INSTRS V)
                                              (RIGHT-INSTRS U))
                                   (WALK (RIGHT-INSTRS V) (RIGHT M)))))
                 (EQUAL (WALK (MAKE-WALK M U V) (WALK U M))
                        (WALK (MAKE-WALK M V U) (WALK V M)))).
```

This simplifies, applying WALK-LAMBDA and WALK-SUBST, and expanding the definitions of COMMAND, AND, MAKE-WALK, EQUAL, and WALK, to two new goals:

```
Case 4.2.
        (IMPLIES (AND (COMBP M)
                      (EQUAL (CADDR U) 'REDUCE)
                      (LAMBDAP (LEFT M))
                      (EQUAL (WALK (MAKE-WALK (LEFT M)
                                              (LEFT-INSTRS U)
                                              (LEFT-INSTRS V))
                                   (WALK (LEFT-INSTRS U) (LEFT M)))
```

```
                (WALK (MAKE-WALK (LEFT M)
                                (LEFT-INSTRS V)
                                (LEFT-INSTRS U))
                      (WALK (LEFT-INSTRS V) (LEFT M))))
          (EQUAL (WALK (MAKE-WALK (RIGHT M)
                                  (RIGHT-INSTRS U)
                                  (RIGHT-INSTRS V))
                       (WALK (RIGHT-INSTRS U) (RIGHT M)))
                 (WALK (MAKE-WALK (RIGHT M)
                                  (RIGHT-INSTRS V)
                                  (RIGHT-INSTRS U))
                       (WALK (RIGHT-INSTRS V) (RIGHT M))))
          (NOT (EQUAL (CADDR V) 'REDUCE)))
     (EQUAL (SUBST (WALK (MAKE-WALK (LEFT M)
                                    (LEFT-INSTRS U)
                                    (LEFT-INSTRS V))
                         (WALK (LEFT-INSTRS U)
                               (BODY (LEFT M))))
                   (WALK (MAKE-WALK (RIGHT M)
                                    (RIGHT-INSTRS U)
                                    (RIGHT-INSTRS V))
                         (WALK (RIGHT-INSTRS U) (RIGHT M)))
                   1)
            (WALK (CONS (MAKE-WALK (LEFT M)
                                   (LEFT-INSTRS V)
                                   (LEFT-INSTRS U))
                        (CONS (MAKE-WALK (RIGHT M)
                                         (RIGHT-INSTRS V)
                                         (RIGHT-INSTRS U))
                              '(REDUCE)))
                  (COMB (WALK (LEFT-INSTRS V) (LEFT M))
                        (WALK (RIGHT-INSTRS V) (RIGHT M)))))))),
```

which again simplifies, rewriting with WALK-LAMBDA, LAMBDA-WALK,
RIGHT-COMB, CAR-CONS, LEFT-COMB, and CDR-CONS, and opening up WALK,
LEFT-INSTRS, RIGHT-INSTRS, EQUAL, COMMAND, and CAR, to:

```
     (IMPLIES (AND (COMBP M)
                   (EQUAL (CADDR U) 'REDUCE)
                   (LAMBDAP (LEFT M))
                   (EQUAL (LAMBDA (WALK (MAKE-WALK (LEFT M)
                                                   (LEFT-INSTRS U)
                                                   (LEFT-INSTRS V))
                                        (WALK (CAR U) (BODY (LEFT M)))))
                          (WALK (MAKE-WALK (LEFT M)
                                           (LEFT-INSTRS V)
                                           (LEFT-INSTRS U))
                                (WALK (LEFT-INSTRS V) (LEFT M))))
                   (EQUAL (WALK (MAKE-WALK (RIGHT M)
                                           (RIGHT-INSTRS U)
                                           (RIGHT-INSTRS V))
                                (WALK (RIGHT-INSTRS U) (RIGHT M)))
                          (WALK (MAKE-WALK (RIGHT M)
                                           (RIGHT-INSTRS V)
                                           (RIGHT-INSTRS U))
                                (WALK (RIGHT-INSTRS V) (RIGHT M))))
                   (NOT (EQUAL (CADDR V) 'REDUCE)))
```

```
                (EQUAL (SUBST (WALK (MAKE-WALK (LEFT M)
                                              (LEFT-INSTRS U)
                                              (LEFT-INSTRS V))
                                   (WALK (CAR U) (BODY (LEFT M))))
                             (WALK (MAKE-WALK (RIGHT M)
                                              (RIGHT-INSTRS U)
                                              (RIGHT-INSTRS V))
                                   (WALK (RIGHT-INSTRS U) (RIGHT M)))
                       1)
                      (SUBST (WALK (MAKE-WALK (LEFT M)
                                              (LEFT-INSTRS V)
                                              (LEFT-INSTRS U))
                                   (WALK (CAR V) (BODY (LEFT M))))
                             (WALK (MAKE-WALK (RIGHT M)
                                              (RIGHT-INSTRS U)
                                              (RIGHT-INSTRS V))
                                   (WALK (RIGHT-INSTRS U) (RIGHT M)))
                       1))),
```

which further simplifies, applying BODY-LAMBDA and LAMBDA-EQUAL, and
opening up the functions LEFT-INSTRS, WALK, and RIGHT-INSTRS, to:

     T.

Case 4.1.
```
      (IMPLIES
            (AND (COMBP M)
                 (EQUAL (CADDR U) 'REDUCE)
                 (LAMBDAP (LEFT M))
                 (EQUAL (WALK (MAKE-WALK (LEFT M)
                                         (LEFT-INSTRS U)
                                         (LEFT-INSTRS V))
                              (WALK (LEFT-INSTRS U) (LEFT M)))
                        (WALK (MAKE-WALK (LEFT M)
                                         (LEFT-INSTRS V)
                                         (LEFT-INSTRS U))
                              (WALK (LEFT-INSTRS V) (LEFT M))))
                 (EQUAL (WALK (MAKE-WALK (RIGHT M)
                                         (RIGHT-INSTRS U)
                                         (RIGHT-INSTRS V))
                              (WALK (RIGHT-INSTRS U) (RIGHT M)))
                        (WALK (MAKE-WALK (RIGHT M)
                                         (RIGHT-INSTRS V)
                                         (RIGHT-INSTRS U))
                              (WALK (RIGHT-INSTRS V) (RIGHT M))))
                 (EQUAL (CADDR V) 'REDUCE))
            (EQUAL (SUBST (WALK (MAKE-WALK (LEFT M)
                                           (LEFT-INSTRS U)
                                           (LEFT-INSTRS V))
                                (WALK (LEFT-INSTRS U)
                                      (BODY (LEFT M))))
                         (WALK (MAKE-WALK (RIGHT M)
                                          (RIGHT-INSTRS U)
                                          (RIGHT-INSTRS V))
                               (WALK (RIGHT-INSTRS U) (RIGHT M)))
                   1)
                  (WALK (SUB-WALK (MAKE-WALK (LEFT M)
```

```
                                   (LEFT-INSTRS V)
                                   (LEFT-INSTRS U))
                       (WALK (LEFT-INSTRS V) (BODY (LEFT M)))
                       (MAKE-WALK (RIGHT M)
                                  (RIGHT-INSTRS V)
                                  (RIGHT-INSTRS U))
             1)
      (SUBST (WALK (LEFT-INSTRS V) (BODY (LEFT M)))
             (WALK (RIGHT-INSTRS V) (RIGHT M))
             1)))).
```

This again simplifies, rewriting with WALK-LAMBDA, LAMBDA-WALK, and
WALK-SUBST, and expanding the functions WALK, LEFT-INSTRS, and EQUAL, to
the new conjecture:

```
      (IMPLIES (AND (COMBP M)
                    (EQUAL (CADDR U) 'REDUCE)
                    (LAMBDAP (LEFT M))
                    (EQUAL (LAMBDA (WALK (MAKE-WALK (LEFT M)
                                                   (LEFT-INSTRS U)
                                                   (LEFT-INSTRS V))
                                         (WALK (CAR U) (BODY (LEFT M)))))
                           (WALK (MAKE-WALK (LEFT M)
                                            (LEFT-INSTRS V)
                                            (LEFT-INSTRS U))
                                 (WALK (LEFT-INSTRS V) (LEFT M))))
                    (EQUAL (WALK (MAKE-WALK (RIGHT M)
                                            (RIGHT-INSTRS U)
                                            (RIGHT-INSTRS V))
                                 (WALK (RIGHT-INSTRS U) (RIGHT M)))
                           (WALK (MAKE-WALK (RIGHT M)
                                            (RIGHT-INSTRS V)
                                            (RIGHT-INSTRS U))
                                 (WALK (RIGHT-INSTRS V) (RIGHT M))))
                    (EQUAL (CADDR V) 'REDUCE))
               (EQUAL (SUBST (WALK (MAKE-WALK (LEFT M)
                                              (LEFT-INSTRS U)
                                              (LEFT-INSTRS V))
                                   (WALK (CAR U) (BODY (LEFT M))))
                             (WALK (MAKE-WALK (RIGHT M)
                                              (RIGHT-INSTRS U)
                                              (RIGHT-INSTRS V))
                                   (WALK (RIGHT-INSTRS U) (RIGHT M)))
                             1)
                      (SUBST (WALK (MAKE-WALK (LEFT M)
                                              (LEFT-INSTRS V)
                                              (LEFT-INSTRS U))
                                   (WALK (CAR V) (BODY (LEFT M))))
                             (WALK (MAKE-WALK (RIGHT M)
                                              (RIGHT-INSTRS U)
                                              (RIGHT-INSTRS V))
                                   (WALK (RIGHT-INSTRS U) (RIGHT M)))
                             1))),
```

which further simplifies, applying BODY-LAMBDA and LAMBDA-EQUAL, and
opening up the definitions of LEFT-INSTRS, WALK, and RIGHT-INSTRS, to:

T.

Case 3. (IMPLIES (AND (NOT (LAMBDAP M))
                      (COMBP M)
                      (NOT (AND (EQUAL (COMMAND U) 'REDUCE)
                                (LAMBDAP (LEFT M))))
                      (AND (EQUAL (COMMAND V) 'REDUCE)
                           (LAMBDAP (LEFT M)))
                      (EQUAL (WALK (MAKE-WALK (LEFT M)
                                             (LEFT-INSTRS U)
                                             (LEFT-INSTRS V))
                                  (WALK (LEFT-INSTRS U) (LEFT M)))
                             (WALK (MAKE-WALK (LEFT M)
                                             (LEFT-INSTRS V)
                                             (LEFT-INSTRS U))
                                  (WALK (LEFT-INSTRS V) (LEFT M))))
                      (EQUAL (WALK (MAKE-WALK (RIGHT M)
                                             (RIGHT-INSTRS U)
                                             (RIGHT-INSTRS V))
                                  (WALK (RIGHT-INSTRS U) (RIGHT M)))
                             (WALK (MAKE-WALK (RIGHT M)
                                             (RIGHT-INSTRS V)
                                             (RIGHT-INSTRS U))
                                  (WALK (RIGHT-INSTRS V) (RIGHT M)))))
               (EQUAL (WALK (MAKE-WALK M U V) (WALK U M))
                      (WALK (MAKE-WALK M V U) (WALK V M)))).

This simplifies, applying RIGHT-COMB, WALK-LAMBDA, CAR-CONS, LAMBDA-WALK,
LEFT-COMB, CDR-CONS, and WALK-SUBST, and expanding the definitions of
COMMAND, AND, MAKE-WALK, EQUAL, WALK, RIGHT-INSTRS, LEFT-INSTRS, and CAR, to:

        (IMPLIES (AND (COMBP M)
                      (NOT (EQUAL (CADDR U) 'REDUCE))
                      (EQUAL (CADDR V) 'REDUCE)
                      (LAMBDAP (LEFT M))
                      (EQUAL (WALK (MAKE-WALK (LEFT M)
                                             (LEFT-INSTRS U)
                                             (LEFT-INSTRS V))
                                  (WALK (LEFT-INSTRS U) (LEFT M)))
                             (WALK (MAKE-WALK (LEFT M)
                                             (LEFT-INSTRS V)
                                             (LEFT-INSTRS U))
                                  (WALK (LEFT-INSTRS V) (LEFT M))))
                      (EQUAL (WALK (MAKE-WALK (RIGHT M)
                                             (RIGHT-INSTRS U)
                                             (RIGHT-INSTRS V))
                                  (WALK (RIGHT-INSTRS U) (RIGHT M)))
                             (WALK (MAKE-WALK (RIGHT M)
                                             (RIGHT-INSTRS V)
                                             (RIGHT-INSTRS U))
                                  (WALK (RIGHT-INSTRS V) (RIGHT M))))
                      (EQUAL (SUBST (WALK (MAKE-WALK (LEFT M)
                                                    (LEFT-INSTRS U)
                                                    (LEFT-INSTRS V))
                                         (WALK (CAR U) (BODY (LEFT M))))
                                   (WALK (MAKE-WALK (RIGHT M)
                                                   (RIGHT-INSTRS U)

```
                                           (RIGHT-INSTRS V))
                             (WALK (RIGHT-INSTRS U) (RIGHT M)))
                  1)
            (SUBST (WALK (MAKE-WALK (LEFT M)
                                    (LEFT-INSTRS V)
                                    (LEFT-INSTRS U))
                         (WALK (LEFT-INSTRS V)
                               (BODY (LEFT M))))
                   (WALK (MAKE-WALK (RIGHT M)
                                    (RIGHT-INSTRS U)
                                    (RIGHT-INSTRS V))
                         (WALK (RIGHT-INSTRS U) (RIGHT M)))
                  1))).
```

This again simplifies, applying WALK-LAMBDA and LAMBDA-WALK, and opening up
WALK and LEFT-INSTRS, to:

```
      (IMPLIES (AND (COMBP M)
                    (NOT (EQUAL (CADDR U) 'REDUCE))
                    (EQUAL (CADDR V) 'REDUCE)
                    (LAMBDAP (LEFT M))
                    (EQUAL (WALK (MAKE-WALK (LEFT M)
                                           (LEFT-INSTRS U)
                                           (LEFT-INSTRS V))
                                 (WALK (LEFT-INSTRS U) (LEFT M)))
                           (LAMBDA (WALK (MAKE-WALK (LEFT M)
                                                    (LEFT-INSTRS V)
                                                    (LEFT-INSTRS U))
                                         (WALK (CAR V) (BODY (LEFT M))))))
                    (EQUAL (WALK (MAKE-WALK (RIGHT M)
                                           (RIGHT-INSTRS U)
                                           (RIGHT-INSTRS V))
                                 (WALK (RIGHT-INSTRS U) (RIGHT M)))
                           (WALK (MAKE-WALK (RIGHT M)
                                           (RIGHT-INSTRS V)
                                           (RIGHT-INSTRS U))
                                 (WALK (RIGHT-INSTRS V) (RIGHT M)))))
               (EQUAL (SUBST (WALK (MAKE-WALK (LEFT M)
                                             (LEFT-INSTRS U)
                                             (LEFT-INSTRS V))
                                   (WALK (CAR U) (BODY (LEFT M))))
                             (WALK (MAKE-WALK (RIGHT M)
                                             (RIGHT-INSTRS U)
                                             (RIGHT-INSTRS V))
                                   (WALK (RIGHT-INSTRS U) (RIGHT M)))
                            1)
                      (SUBST (WALK (MAKE-WALK (LEFT M)
                                             (LEFT-INSTRS V)
                                             (LEFT-INSTRS U))
                                   (WALK (CAR V) (BODY (LEFT M))))
                             (WALK (MAKE-WALK (RIGHT M)
                                             (RIGHT-INSTRS U)
                                             (RIGHT-INSTRS V))
                                   (WALK (RIGHT-INSTRS U) (RIGHT M)))
                            1))),
```

which further simplifies, rewriting with BODY-LAMBDA and LAMBDA-EQUAL, and opening up the functions LEFT-INSTRS, WALK, and RIGHT-INSTRS, to:

     T.

```
Case 2. (IMPLIES (AND (NOT (LAMBDAP M))
                      (COMBP M)
                      (NOT (AND (EQUAL (COMMAND U) 'REDUCE)
                                (LAMBDAP (LEFT M))))
                      (NOT (AND (EQUAL (COMMAND V) 'REDUCE)
                                (LAMBDAP (LEFT M))))
                      (EQUAL (WALK (MAKE-WALK (LEFT M)
                                             (LEFT-INSTRS U)
                                             (LEFT-INSTRS V))
                                  (WALK (LEFT-INSTRS U) (LEFT M)))
                             (WALK (MAKE-WALK (LEFT M)
                                             (LEFT-INSTRS V)
                                             (LEFT-INSTRS U))
                                  (WALK (LEFT-INSTRS V) (LEFT M))))
                      (EQUAL (WALK (MAKE-WALK (RIGHT M)
                                             (RIGHT-INSTRS U)
                                             (RIGHT-INSTRS V))
                                  (WALK (RIGHT-INSTRS U) (RIGHT M)))
                             (WALK (MAKE-WALK (RIGHT M)
                                             (RIGHT-INSTRS V)
                                             (RIGHT-INSTRS U))
                                  (WALK (RIGHT-INSTRS V) (RIGHT M)))))
                 (EQUAL (WALK (MAKE-WALK M U V) (WALK U M))
                        (WALK (MAKE-WALK M V U) (WALK V M)))).
```

This simplifies, rewriting with RIGHT-COMB, LEFT-COMB, CAR-CONS, and
CDR-CONS, and opening up the definitions of COMMAND, AND, MAKE-WALK, WALK,
RIGHT-INSTRS, LEFT-INSTRS, EQUAL, and CAR, to:

     T.

```
Case 1. (IMPLIES (AND (NOT (LAMBDAP M))
                      (NOT (COMBP M)))
                 (EQUAL (WALK (MAKE-WALK M U V) (WALK U M))
                        (WALK (MAKE-WALK M V U) (WALK V M)))),
```

which simplifies, opening up the functions MAKE-WALK and WALK, to:

     T.

    That finishes the proof of *1.   Q.E.D.

[0.0 3.0 12.6]
MAIN

6.7 Analysis

In this section we first carry out a quantitative analysis of the proof and follow that up with some qualitative observations. The quantitative analysis compares the machine-readable proof with several published versions of the proof and also with the outline in Sections 6.1 and 6.2 of this chapter. Perhaps the simplest metric that can be defined for a proof is its length in terms of the number of words. This metric was first defined by de Bruijn in the context of the AUTOMATH system [dB80]. It was observed that the ratio between the length of the machine-readable proof and the informal proof was between 10 and 50. This ratio was termed the *loss factor*. It was also observed that the loss factor did not vary significantly from one proof to another. Knoblock and Constable [KC86] label a proof checker as being *linear* if it has a constant loss factor. Based on empirical evidence, the AUTOMATH proof checkers and the Nuprl proof checker [KC86] have been observed to be linear.

We note that:

1. The notions of an *informal proof* and of a *word* in an informal proof cannot be made precise. An informal proof might contain diagrams and employ linguistic devices such as ellipses and these can convey arbitrary amounts of information to the reader.

2. It is not being argued that the length of a proof in terms of the number of words is an appropriate metric for a proof. Since we are comparing two pieces of text in which words are the smallest meaningful units, the above metric seems adequate for the kind of superficial analysis carried out below.

3. The intent of comparing the lengths of the machine-readable proof and the informal proof is not to imply that these are alternatives to one another. Rather it is to rate the effectiveness of a proof checker across a broad spectrum of proofs, and to compare one proof checker with another when they do not both check the same class of proofs.

4. The proof of the Church–Rosser theorem is too small an example from which to make any conclusions about the linearity of the Boyer–Moore theorem prover. This would require a similar analysis of a number of other examples or a systematic mechanical verification of a substantial portion of a standard text.

5. We explain how the numbers were estimated since some of the figures are accurate word counts and others are approximate.

A detailed comparison is made between the informal proof (call it I) that is sketched in Section 6.2 of this chapter and the machine-readable proof (M). We also present figures from the proof given in the text by Hindley and Seldin (HS) [HS86], the proof in Stenlund's book (St) [Ste72] and the one described by de Bruijn (dB) [dB72].

The total length of M is 4624 words. The total length of I is 3476 words. This figure includes all the words from the beginning of Section 6.1.1 to the end of Section 6.2, excluding those in the examples. The loss factor with respect to I is 1.33.

The entire proof HS spans 21 pages at roughly 300 words per page giving a total of approximately 6300 words. The introductory definitions and lemmas span about 13 pages, and the actual proof of the theorem about 8 pages. The loss factor when measured with respect to this proof would be 0.73. It is very rare for a sizable machine-checked proof to be shorter than the corresponding description in a standard text.

The entire proof St consists of 14 pages at around 300 words per page and hence a total of 4200 words. The proof sketch dB spans about 9 pages at approximately 400 words per page yielding a total of 3600 words. These figures indicate that I is a reasonable basis for our comparisons.

Now we analyse each section of the proof. The introduction of the basic notions of lambda calculus sufficient to state the Church–Rosser theorem takes up fewer than 400 words in M. In I the introductory definitions cover 1200 words. This is clearly where the bulk of the efficiency of M is gained.

Next we compare the proofs of the diamond property of walks in M (de Bruijn version) and in I. The comparison seems fair since both of these proofs ignore α-steps. The proof in M is about 750 words. The proof in I is 1200 words. Thus the same steps of reasoning have been described more efficiently to a proof checker than in a sketchy, informal description.

The argument demonstrating that the transitive closure of the *walks to* relation has the diamond property covers around 500 words in M and only 300 words in I.

Showing that the reduction relation is the transitive closure of a walk takes up 600 words in M but is dismissed as being obvious in about 44 words in I.

The remainder of M covers the correspondence between the two notations thus deriving the Church–Rosser theorem for the standard notation. This takes up 2400 words in M and has no counterpart in I.

The main conclusion to be drawn from the above analysis is that a difficult mathematical argument can be efficiently described to an automatic proof checker. We have seen above that some parts of the machine-readable proof are actually shorter than their informal counterparts.

On the qualitative side, the questions examined are:

1. Does this mechanical proof enhance our understanding of the proof of the Church–Rosser theorem?

2. How does the mechanical proof compare qualitatively with the previously published proofs?

3. Does it tell us anything about the capabilities of the Boyer–Moore theorem prover?

4. Is there anything to distinguish this mechanical proof from similar verifications?

5. What insights can be gained into the process of constructing a machine-verifiable proof from a descriptive argument?

To deal with the first of these questions, these mechanical proofs do not contain any new concepts nor do they employ any previously unknown facts about the lambda calculus. The intention of the project was to verify a proof similar to the Tait–Martin-Löf proof, and not to discover a new proof. The primary motivation for such a verification was that the theorem had a long history of candidate proofs that were later found to be inadequate. The difficulty was perhaps due to the combinatorial explosion inherent in the notion of a β-step and the problems posed by substitution in the presence of bound variables. The de Bruijn representation solves some of the problems with substitution and α-equivalence, but the notation is not easy to intuitively grasp. Initially many of the operations on de Bruijn terms were defined inelegantly or incorrectly and many of the lemmas were posed incorrectly. These mistakes and inelegances were quickly exposed by the attempted mechanical proofs. So the main contribution is a carefully crafted sequence of definitions and lemmas which make the logical structure and the elegance of the Tait–Martin-Löf argument apparent, and greatly enhances our confidence in its correctness. An examination of the proof script also reveals that the proof was carried out within a subset of the Boyer–Moore logic which is essentially primitive recursive arithmetic.

The mechanical proof is structured in the following sequence:

1. The formalization of the lambda calculus.

2. The diamond property of walks in the de Bruijn representation.

3. The Church–Rosser theorem (de Bruijn).

4. Proving that a reduction is the transitive closure of a walk.

5. Correspondence between the de Bruijn and standard notations.

The above sequence makes it possible to present the core of the argument more economically than previous descriptions. The mechanical proof makes clear exactly which properties of substitution are needed to prove the diamond property of walks. The proof here is also a notational improvement over de Bruijn's presentation [dB72] where substitution is defined using an infinite association list of variable–term pairs. The mechanical proof loses some of its cogency in the "tricky" details of "obvious" steps such as steps 4 and 5 above. The radically different views of terms and substitution in the de Bruijn and standard notations made step 5 difficult to carry out using the theorem prover.

These experiments tell us a great deal about the Boyer–Moore theorem prover and about proof checkers in general. The theorem prover's principle of definition makes it easy to describe in a concise and natural way, the basic concepts required to formulate the theorem. The proofs demonstrate how the power and sophistication of the Boyer–Moore induction heuristic pays off in a big way. The proofs of several of the lemmas that were proved automatically would tax most humans. In many cases the proof was conveyed more efficiently to the theorem prover than it could have been conveyed to another trained human. The structured manner in which the theorem prover proceeds about a proof makes it significantly easier to construct a counterexample from a failed proof attempt. The facility that the theorem prover has to carry out definitions by recursion and proofs by induction, and to exploit the duality between recursion and induction has been especially useful in doing proofs in metamathematics. In number theory, for example, the appropriate induction scheme is less likely to be implicitly available in the definitions of functions appearing in a given conjecture.

The key insight into the process of constructing proofs which will be mechanically checked is that the form of a representation, a definition or a conjecture can significantly affect the efficiency of the proof. The first attempt to prove the Church–Rosser theorem quickly ran into problems due to free–bound variable complications. The attempt to carry out the proof with the de Bruijn notation seemed similarly headed until the substitution operation SUBST was defined as in Section 6.4.2. From that point to the verification of the diamond property of walks, which is the core of the proof, took merely six hours of work with the theorem prover. The fact that SUBST is defined to maintain a simple invariant makes it easy to state and prove theorems about substitution. There were other occasions in the proof when minor changes in the form of a definition or representation significantly improved the felicity with which theorems could be stated and proofs described to the theorem prover.

As an illustration of the clearer understanding gained from the process of machine checking a proof, a verification of the diamond property of walks was carried out directly in the standard notation as a postscript to the above proof. This was the proof that was originally attempted and the free–bound variable confusion was eventually resolved employing some of the insights gained from working with the de Bruijn notation.

Chapter 7

Conclusions

*If the mathematical process were really one
of strict, logical progression, we would
still be counting on our fingers.*
R. A. De Millo, R. J. Lipton, A. J. Perlis [MLP79]

Specific conclusions associated with individual proofs have already been presented in the precedings chapters. In this chapter, we state some general conclusions and advance some specific comments on the Boyer–Moore theorem prover.

The social process by which mathematical arguments are scrutinized and accepted or rejected remains important even with the introduction of mechanized proof checking. The paper [MLP79] that we have quoted above, argues that mathematical proofs are interesting to mathematicians and are therefore examined with great care by means of the social process. The authors contend that proofs of computer programs are unlikely to be of similar interest to the social process. They express skepticism as to whether machine-assisted verification of programs and proofs could ever be feasible or illuminating. Though these arguments carry considerable weight, they are founded on some serious misconceptions on the role of proofs, formal proofs, and of mechanical verification. Proofs are not a means to obtaining certitude. As Lakatos [Lak76, page 48] writes, "the virtue of a logical proof is not that it compels belief, but that it suggests doubts." While formalism might not always serve as a mechanism for the discovery of proofs or counterexamples, it can be used to expose the weaknesses in an argument and to isolate the assumptions on which an argument rests. Formalism is also useful in observing (and mechanizing) patterns of proofs. Formalization and mechanization enjoy a synergistic relationship: formalism makes it possible to represent mathematics in a mechanized form, and mechanization makes it feasible to apply formalism with the sharpest possible rigor.

The unaided social process has its drawbacks. Critics of mechanized verification have argued that mechanization subverts the social process by which purported proofs come to be discarded or accepted [MLP79]. The social process is of course enormously important, but it has not proved very effective at detecting and filtering out erroneous proofs. For example, the interactive convergence algorithm for Byzantine fault-tolerant clock synchronization due to Lamport and Melliar-Smith [LMS85] appears in a landmark journal paper that was published in 1985. The mechanical verification of this algorithm in 1988 by Rushby and von Henke [RvH91] (using EHDM [RvHO91]) revealed that four out of five lemmas and

183

the proof of the main theorem were false, and none of these errors was identified by the intervening social process.

An automatic proof checker is not a tool for certifying that proofs are correct. Only a very small percentage of the time spent with a proof checker is devoted to correctly proving correct theorems. The overwhelming fraction of time and effort is devoted to discovering mistakes in definitions, theorems, and proofs. All this can be done without the aid of a computer, but few humans have the concentration or attention span needed to conduct proof checking without mechanical assistance.

Should we regard a mechanically verified proof as certifiably correct? Definitely not. As mentioned above, mechanical assistance is useful in finding and correcting reasoning errors but it cannot guarantee that errors have been eliminated. For instance, it is quite possible that the theorems we have formally proved in the previous chapters have been misstated and do not capture the content of their original statements. It is also possible to introduce errors in the informal application of formal definitions and theorems.

Proof checking is a creative activity. It would be a serious misconception to regard the process of formalizing and verifying an informal mathematical argument as an unimaginative derivation of the conclusion from the axioms and definitions. The formal process is extremely sensitive to the manner in which concepts are represented or defined. Representational details such as the choice of notations, the use of functions *versus* predicates and sets *versus* lists, have a profound effect on the ease of formalization and proof. For example, the use of de Bruijn indices to represent bound variables significantly simplifies both the informal and the mechanical proofs of the Church–Rosser theorem.

Mechanical verification can complement the social process. The process of mechanical verification is a creative one and yields important insights that might be easily missed by the unaided intellect. The metamathematical proofs presented in this proof have been thoroughly studied in the literature, and for this reason, no specific errors were found. Even so, the mechanical verification did yield a more rigorous argument than other presentations of similar proofs without any significant loss of cogency. Proofs involving the correctness of digital systems tend to contain a lot of detail but these proofs usually follow certain systematic patterns. Here, unlike traditional mathematical proofs, there often is no single flash of mathematical insight that makes it easy to comprehend all the details of the argument. Mechanization not only makes it much simpler to generate such proofs in a systematic manner but also helps to identify the needed assumptions, and points the way to generalizations and improvements.

The Boyer–Moore prover generates an intelligible commentary explaining the reasoning underlying its attempted proof. When proof attempts failed, as they often do, this commentary can be examined to pinpoint the source of the failure, discover counterexamples, and to repair the definitions and theorems as needed. Even when proof attempts succeed, the proof could be further improved based on an examination of the commentary generated by the theorem prover.

Are the proofs described here convincing, surveyable, and formalizable? Tymoczko [Tym79] gave these criteria for a reasonable proof in order to show that the computer proof of the four-color theorem [AH76] did not qualify as such a proof. The proofs here are qualitatively different from that of the four-color theorem. In our proofs, a computer is used to check the deduction steps whereas in the case of the four-color theorem, the computer was used to carry out a large case analysis by calculation. The computer-assisted proof of the four-color theorem, though convincing, is not, in this sense, surveyable, whereas the proofs here are, in principle, surveyable even without mechanical assistance. It is left to the reader to decide whether the proofs presented here are convincing, but they clearly are formalizable.

A mechanization of metamathematical proofs can be useful to automated proof checking. As the proofs in the preceding chapter demonstrate, metatheorems can be used as a labor-saving device in order to simplify the task of proof construction. A metamathematically extensible proof checker [DS77] is one that is capable of proving and exploiting such metatheorems. We have already seen how the metatheorems and proofs presented in Chapter 3 about tautologies, instantiations, equalities, and equivalences were useful in extending the proof checking capabilities of the proof checker PROVES. Howe [How88] describes the introduction of metamathematical extensibility into Nuprl by means of partial computational reflection. There are very few other such efforts to make proof checkers extensible; this is clearly a research area with considerable promise.

The interaction between metamathematics and automated deduction poses several important and interesting challenges. The first challenge, as just mentioned, is that of proving and exploiting metatheorems within the framework of an automated proof checker. We have employed a theory of recursive Lisp functions as the metatheoretical framework for the present work. There clearly are a number of other candidate metatheoretical frameworks that are of interest. The various logical frameworks [HHP87, MNPS91, Pau90, Fel89, Pfe89] are convenient for defining logics within the framework of a typed lambda calculus or a higher-order logic programming language. The use of such frameworks for proving useful metatheorems remains an open challenge. Feferman [Fef88] has proposed a theory of recursive functions and inductively defined classes as a metatheoretic framework in which logics can be defined and metatheorems proved. Feferman's theory is closely related to the pure Lisp formalism used by the Boyer–Moore theorem prover.

There are also several interesting metatheorems that have not yet been mechanically verified. The second incompleteness theorem asserts that no theory such as Z2 can prove a statement asserting its own consistency. One way of proving it is to show that the metatheoretic argument used to prove the first incompleteness theorem can be formalized within Z2 itself. If Z2 could prove its own consistency, then the formalized incompleteness theorem could be used to show that Z2 would be able to prove the statement asserting the unprovability of the sentence U. Since U is the statement asserting its own unprovability, we would have an inconsistency in Z2. The proof of the second incompleteness theorem was not attempted by us. Its proof as sketched above would be quite tedious to formalize, but it is possible that other alternative proofs might be more easily verified. A mechanical proof of the independence of the continuum hypothesis is another significant challenge. Cantor's continuum hypothesis asserted that the first uncountable cardinal was the cardinality of the real numbers. Gödel [Göd40] showed that set theory (with the axiom of choice) when extended with this axiom was consistent relative to the unextended set theory, so that the continuum hypothesis could not be disproved in the unextended set theory assuming the consistency of the latter. Later, Cohen [Coh66] showed that continuum hypothesis could also not be disproved in set theory, thus making it an undecidable sentence for set theory. Paris and Harrington's demonstration [PH78] of a natural undecidable statement of Peano arithmetic is yet another worthy candidate for mechanized verification.

We now discuss some specific conclusions about the Boyer–Moore theorem prover.

7.1 A Critique of the Boyer–Moore Theorem Prover

The following aspects of the Boyer–Moore theorem prover were crucial to the success of these proofs:

1. The Lisp based language of the Boyer–Moore logic is expressive enough to facilitate relatively direct formulations of many of the definitions and theorems of metamathematics.

2. The theorem prover is sophisticated enough to perform the low-level deductions (and some of the high-level deductions) autonomously through its use of heuristics for simplification, rewriting, and well-founded induction.

3. Despite this sophistication, a trained user can direct the theorem prover to check a desired proof rather than to merely establish that a given assertion is a theorem.

4. The theorem prover generates as output a well-structured description of its attempted proof which makes it possible to uncover errors in the formulation, and to check on the progress of the proof.

5. By introducing "break-points", it is possible to track the application of rewrite rules in a proof. Mistakes in the statement of the conjecture are frequently located at one of these "break-points".

The Boyer–Moore theorem prover has acquired some new features since this work was done. There is now a LET macro that allows frequently used subexpressions to be given a name. This feature would have been quite useful to have for our proofs. There is a definition facility to introduce functions that are not defined but merely *constrained* by means of an axiom, provided an instantiation can be exhibited for the axiom. There is also some limited capability for handling first-order quantification. Kaufmann [Kau88] has implemented some proof checking extensions to the system. It is not clear what impact these changes to the theorem prover might have on the proofs of the incompleteness theorem and the Church–Rosser theorem.

The following are some of the drawbacks of the Boyer–Moore theorem prover:

1. The shell principle is quite rigid and does not permit, for example, the addition of a shell corresponding to the well-formed formulas of first-order logic.

2. The absence of quantifiers renders the statements of the lemmas somewhat unnatural from the viewpoint of a mathematician. The statement of the incompleteness theorem, for example, is in appearance different from the standard formulation of the theorem.

3. The induction heuristic is not sophisticated enough, and in many cases the appropriate induction scheme had to be provided as a hint to the theorem prover. All of the induction schemes in the proofs of the computability of the metatheoretic functions by the Lisp interpreter were supplied as hints.

4. The facility to extend the theorem prover through simplifiers (termed *metafunctions*) turned out to be too restricted. It was not possible, for instance, to prove the computability of all of the metatheoretic functions by the Lisp interpreter, as a theorem in the metatheory of the Boyer–Moore logic.

5. The theorem prover does not permit the use of rewrite rules for congruence relations other than the strict equality predicate, EQUAL. This proved to be a drawback since it was not possible to replace α-equivalent subterms by one another by means of rewrite rules. There have been some recent attempts to allow rewriting relative to congruence relations, but the status of this work is still unclear.

6. The theorem prover is not very flexible in allowing definitions or theorems to be modified. Even minor modifications to previously introduced definitions or theorems require a significant amount of work.

7. It was frequently necessary to enable and disable function definitions and rewrite rules in the course of the proof so that these definitions and rules could be used in some contexts but not in others. It is a fairly tedious trial-and-error process to determine which definitions and lemmas should be enabled or disabled.

8. There is a facility for providing hints to the theorem prover to employ a particular form of induction, to enable or disable certain definitions or lemmas, or to use lemmas in a certain way. This facility is somewhat clumsy to use since a user must make a number of guesses about the form of the mechanical proof. It also makes the proof less elegant since the hinted lemmas are instantiated and added as hypotheses to the conjectured theorem.

Overall, the convenience of a powerful heuristic theorem prover such as the Boyer–Moore theorem prover is that a number of trivial, and some fairly nontrivial, theorems can be proved with minimal effort. The price for this convenience is that it is very difficult to control such systems when their heuristics fail. There really are no fully automatic theorem provers; the user has to somehow convey the proof to the theorem prover. The more mechanized theorem provers expect the proof to be supplied through the specific forms of the definitions or lemmas, through various settings of flags and parameters, and through hints given with the theorem statement. The human directives to these systems often have little mathematical content. In interactive proof checking systems, on the other hand, the proof is conveyed at a lower level of granularity, but the human input usually carries some mathematical meaning. The crucial challenge for automated reasoning is to find ways of combining the advantages of mechanized theorem provers and interactive proof checkers. In the end, automated proof construction must be an activity that enhances the quality and impact of the social process.

7.2 Summary

The Boyer–Moore theorem prover is a computer program that can be used to prove theorems about Lisp functions. Lisp functions can be used to represent metamathematics. We have used the Boyer–Moore theorem prover to prove metatheorems such as the tautology theorem, Gödel's incompleteness theorem, and the Church–Rosser theorem. Some aspects of these proofs were quite easy to verify while others turned out to be quite difficult. These exercises suggest various ways in which the technology of mechanized theorem proving and proof checking can be further improved. Automated reasoning technology is still in its infancy. When this technology is more fully developed, it is quite conceivable that computers will be widely regarded and used as deductive as well as computational engines.

Bibliography

[AH76]   K. I. Appel and W. Haken. Every planar map is four colorable. *Bull. Am. Math. Soc.*, 82:711–712, 1976.

[AMCP84]   P. B. Andrews, D. A. Miller, E. L. Cohen, and F. Pfenning. Automating higher-order logic. In W. W. Bledsoe and D. W. Loveland, editors, *Automated Theorem Proving: After 25 Years*, pages 169–192. American Mathematical Society, Providence, R.I., 1984.

[Bar78a]   H. P. Barendregt. *The Lambda Calculus, its Syntax and Semantics*. North-Holland, Amsterdam, 1978.

[Bar78b]   J. Barwise. First-order logic. In *Handbook of Mathematical Logic* [Bar78c], pages 5–46.

[Bar78c]   J. Barwise, editor. *Handbook of Mathematical Logic*, volume 90 of *Studies in Logic and the Foundations of Mathematics*. North-Holland, Amsterdam, 1978.

[Ber26]   P. Bernays. Axiomatische untersuchung des aussagen-kalkuls der "Principia mathematica". *Mathematische Zeitschrift*, 25:305–320, 1926.

[BHMY89]   W. R. Bevier, W. A. Hunt, J. S. Moore, and W. D. Young. Special issue on system verification. *Journal of Automated Reasoning*, 5(4):409–530, 1989.

[BJ89]   G. S. Boolos and R. C. Jeffrey. *Computability and Logic, 3rd edition*. Cambridge Univ. Press, 1989.

[Ble77]   W. W. Bledsoe. Non-resolution theorem proving. *Artificial Intelligence*, 9:1–36, 1977.

[Ble83]   W. W. Bledsoe. The UT prover. Technical report ATP-17B, University of Texas at Austin, 1983.

[BM79]   R. S. Boyer and J. S. Moore. *A Computational Logic*. Academic Press, New York, NY, 1979.

[BM81]   R. S. Boyer and J. S. Moore. Metafunctions: Proving them correct and using them efficiently as new proof procedures. In R. S. Boyer and J. S. Moore, editors, *The Correctness Problem in Computer Science*. Academic Press, New York, NY, 1981.

[BM84a]   R. S. Boyer and J. S. Moore. A mechanical proof of the Turing completeness of pure Lisp. *Contemporary Mathematics*, 29:133–167, 1984.

[BM84b]   R. S. Boyer and J. S. Moore. A mechanical proof of the unsolvability of the halting problem. *JACM*, 31(3):441–458, 1984.

[BM84c]   R. S. Boyer and J. S. Moore. Proof checking the RSA public key encryption algorithm. *American Mathematical Monthly*, 91(3):181–189, 1984.

[BM88]   R. S. Boyer and J. S. Moore. *A Computational Logic Handbook*. Academic Press, New York, NY, 1988.

[Boo54]   G. Boole. *An Investigation of the Laws of Thought*. London, 1854. Republished by Dover, New York, 1958.

[Bou68]   N. Bourbaki. *Theory of Sets*. Elements of Mathematics. Addison-Wesley, Reading, MA, 1968.

[BP83]   Paul Benacerraf and Hilary Putnam, editors. *Philosophy of Mathematics: Selected Readings*. Cambridge University Press, Cambridge, England, second edition, 1983.

[Bro80]   F. M. Brown. An investigation into the goals of research in automatic theorem proving as related to mathematical reasoning. *Artificial Intelligence*, 14(3):221–242, October 1980.

[Can55]   G. Cantor. *Contributions to the Founding of the Theory of Transfinite Numbers*. Dover, New York, NY, 1955. Translated and with an 80 page introduction by Philip E. B. Jourdain. Originally published as *Beiträge zur Begründung der Transfiniten Mengenlehre*, 1895 and 1897.

[CH85]   T. Coquand and G. P. Huet. Constructions: A higher order proof system for mechanizing mathematics. In *Proceedings of EUROCAL 85, Linz (Austria)*, Berlin, 1985. Springer-Verlag.

[Chu36]   A. Church. An unsolvable problem of elementary number theory. *American Journal of Mathematics*, 58:345–363, 1936. Reprinted in [Dav65].

[Chu41]   A. Church. *The Calculi of Lambda-Conversion*. Princeton University Press, Princeton, NJ, 1941.

[CL73]   C. L. Chang and R. C. T. Lee. *Symbolic Logic and Mechanical Theorem Proving*. Academic Press, New York, NY, 1973.

[Coh66]   P. J. Cohen. *Set Theory and the Continuum Hypothesis*. Benjamin, New York, NY, 1966.

[Con86]   R. L. Constable, *et al*. *Implementing Mathematics with the Nuprl*. Prentice-Hall, New Jersey, 1986.

[CR36]   A. Church and J. B. Rosser. Some properties of conversion. *Trans. Amer. Math. Soc.*, 39:472–482, 1936.

[Dav65]   M. Davis, editor. *The Undecidable*. Raven Press, Hewlett, N.Y., 1965.

[dB70]   N. G. de Bruijn. The mathematical language AUTOMATH, its usage and some of its extensions. In *Symposium on Automatic Demonstration*, volume 125 of *Lecture Notes in Mathematics*, pages 29–61. Springer-Verlag, Berlin, 1970.

[dB72]   N. G. de Bruijn. Lambda calculus notation with nameless dummies, a tool for automatic formula manipulation, with application to the Church-Rosser theorem. *Indag. Math.*, 34(5):381–392, 1972.

[dB80]   N. G. de Bruijn. A survey of the project AUTOMATH. In J. P. Seldin and J. R. Hindley, editors, *Essays on Combinatory Logic, Lambda Calculus and Formalism*, pages 589–606. Academic Press, New York, NY, 1980.

[Ded63]   R. Dedekind. *Essays on the Theory of Numbers*. Dover, New York, NY, 1963. Reprint of the 1901 translation by Wooster Woodruff Beman; the original essay entitled *Was sind und was sollen die Zahlen?* was written 1888.

[DS77]   M. Davis and J. Schwartz. Metamathematical extensibility for theorem verifiers and proof-checkers. Courant Computer Science Report 12, Courant Institute of Mathematical Sciences, New York, 1977.

[End72]   H. B. Enderton. *A Mathematical Introduction to Logic*. Academic Press, New York, NY, 1972.

[Euc56]   Euclid. *The Thirteen Books of The Elements*. Dover, New York, NY, 1956. Translated with introduction and commentary by Sir Thomas L. Heath.

[Fef82]   S. Feferman. Inductively presented systems and the formalization of metamathematics. In D. van Dalen, D. Lascar, and J. Smiley, editors, *Logic Colloquium '80*. North-Holland, Amsterdam, 1982.

[Fef88]   S. Feferman. Finitary inductively presented logics. In *Logic Colloquium '88*, pages 191–220, Amsterdam, 1988. North-Holland.

[Fel89]   A. Felty. *Specifying and Implementing Theorem Provers in a Higher-Order Logic Programming Language*. PhD thesis, University of Pennsylvania, 1989.

[Fet88]   J. H. Fetzer. Program verification: The very idea. *Communications of the ACM*, 31(9):1048–1063, September 1988.

[FJWDK+86] S. Feferman, Jr. J. W. Dawson, S. C. Kleene, G. H. Moore, R. M. Solovay, and J. van Heijenoort, editors. *Kurt Gödel: Collected Works Vol. 1*. Oxford University Press, 1986.

[Fre67]   G. Frege. Begriffsschrift, a formula language, modeled upon that of arithmetic, for pure thought. In van Heijenoort [vH67], pages 1–82. First published 1879.

[GB80]   P. T. Geach and M. Black, editors. *Translations from the Philosophical Writings of Gottlob Frege*. Rowman and Littlefield, Totowa, New Jersey, third edition, 1980.

[Gen69]   G. Gentzen. The consistency of elementary number theory. In M. E. Szabo, editor, *The Collected Papers of Gerhard Gentzen*, pages 132–213. North-Holland, 1969.

[GMW79]   M. J. Gordon, A. J. Milner, and C. P. Wadsworth. *Edinburgh LCF*, volume 78 of *Lecture Notes in Computer Science*. Springer-Verlag, Berlin, 1979.

[Göd40]   K. Gödel. *The Consistency of the Continuum Hypothesis*. Princeton University Press, 1940.

[Göd44]   K. Gödel. Russell's mathematical logic. In P. A. Schilpp, editor, *The Philosophy of Bertrand Russell*, pages 125–153. Northwestern U. Press, Evanston, Illinois, 1944. Reprinted in [BP83, pages 447–469].

[Göd67a]   K. Gödel. The completeness of the axioms of the functional calculus of logic. In van Heijenoort [vH67], pages 582–591. First published 1930.

[Göd67b]   K. Gödel. On formally undecidable propositions of *Principia mathematica* and related systems. In van Heijenoort [vH67], pages 596–616. Originally published in German in 1931. Translated into English by Jean van Heijenoort. Reprinted in [FJWDK+86, pages 145–195] and [Sha88b, pages 17–47]. Other English translations appear in [Dav65, pages 5–38] and [Göd92].

[Göd92]   Kurt Gödel. *On Formally Undecidable Propositions of Principia Mathematica and Related Systems*. Dover Publications, Inc., New York, NY, 1992. Translated by B. Meltzer, with an Introduction by R. B. Braithwaite. Originally published as a book in 1962. Article originally published in 1931.

[Goo64]   R. L. Goodstein. *Recursive Number Theory*. North-Holland, Amsterdam, 1964.

[Gor88]   M. J. C. Gordon. HOL: A proof generating system for higher-order logic. In G. Birtwistle and P. A. Subrahmanyam, editors, *VLSI Specification, Verification and Synthesis*, pages 73–128. Kluwer, Dordrecht, The Netherlands, 1988.

[Her67]   J. Herbrand. Investigations in proof theory. In van Heijenoort [vH67], pages 525–581. First published 1930.

[HHP87]   R. Harper, F. Honsell, and G. D. Plotkin. A framework for defining logics. In *IEEE Symposium on Logic in Computer Science*, Ithaca, NY, 1987.

[HLS72]   J. R. Hindley, B. Lercher, and J. P. Seldin. *Introduction to Combinatory Logic*. Cambridge University Press, London, 1972.

[How88]   D. J. Howe. Computational metatheory in Nuprl. In E. Lusk and R. Overbeek, editors, *Proceedings of CADE-9*, volume 310 of *Lecture Notes in Computer Science*, pages 238–257, Berlin, 1988. Springer-Verlag.

[HS86]   J. R. Hindley and J. P. Seldin. *Introduction to Combinators and λ-Calculus*. Cambridge Univ. Press, 1986.

[Hun85]   W. A. Hunt, Jr. FM8501: A verified microprocessor. Technical Report 47, University of Texas at Austin, Institute for Computing Science, December 1985.

[Kau88]   M. Kaufmann. A user's manual for an interactive enhancement to the Boyer-Moore theorem prover. Technical Report 19, Computational Logic Inc., Austin, Texas, 1988.

[KC86]   T. B. Knoblock and R. L. Constable. Formalizing metareasoning in type theory. In *IEEE Symposium on Logic in Computer Science*, Cambridge, MA, 1986.

[Kle52]   S. C. Kleene. *Introduction to Metamathematics*. North-Holland, Amsterdam, 1952.

[KZ89]   D. Kapur and H. Zhang. RRL: Rewrite rule laboratory. Technical report, Department of Computer Science, University of Iowa, 1989.

[Lak76]   I. Lakatos. *Proofs and Refutations*. Cambridge University Press, Cambridge, England, 1976.

[Lei65]   G. W. Leibniz. On the universal science: Characteristic. In *Monadology and Other Philosophical Essays*, pages 11–21. The Library of Liberal Arts, 1965. Translated by Paul Schrecker and Anne-Martin Schrecker.

[LH85]   C. Lengauer and C.-H. Huang. The static derivation of concurrency and its mechanized certification. In S. D. Brookes, A. W. Roscoe, and G. Winskel, editors, *Seminar on Concurrency*, volume 197 of *Lecture Notes in Computer Science*, pages 131–150. Springer-Verlag, Berlin, 1985.

[LMS85]   L. Lamport and P. M. Melliar-Smith. Synchronizing clocks in the presence of faults. *Journal of the ACM*, 32(1):52–78, January 1985.

[MAE⁺65]   J. McCarthy, P. W. Abrahams, D. J. Edwards, T. P. Hart, and M. I. Levin. *LISP 1.5 Programmer's Manual*. The MIT Press, Cambridge, MA, 1965.

[Mar84]   R. L. Martin, editor. *Recent Essays on Truth and the Liar Paradox*. Oxford University Press, Oxford, 1984.

[Mas86]   I. A. Mason. *The Semantics of Destructive Lisp*. CSLI Lecture Notes Number 5. CSLI, Stanford, California, 1986.

[McC60]   J. McCarthy. Recursive functions of symbolic expressions and their compu-
           tation by machine. *Communications of the ACM*, 3(4):184–195, 1960.

[McC62]   J. McCarthy. Computer programs for checking mathematical proofs. In *Re-
           cursive Function Theory, Proceedings of a Symposium in Pure Mathematics*,
           volume V, pages 219–227, Providence, Rhode Island, 1962. American Math-
           ematical Society.

[McC63]   J. McCarthy. A basis for a mathematical theory of computation. In P. Braf-
           fort and D. Hershberg, editors, *Computer Programming and Formal Systems*.
           North-Holland, 1963.

[McC78]   J. McCarthy. History of LISP. *ACM SIGPLAN Notices*, 13(8), 1978.

[McC90]   W. McCune. OTTER 2.0 users guide. Technical Report ANL-90/9, Argonne
           National Laboratory, 1990.

[Min67]   M. L. Minsky. *Computation: Finite and Infinite Machines*. Prentice-Hall,
           New Jersey, 1967.

[MJ84]   F. L. Morris and C. B. Jones. An early program proof by Alan Turing. *Annals
           of the History of Computing*, 6:139–143, 1984.

[MLP79]   R. A. De Millo, R. J. Lipton, and A. J. Perlis. Social processes and proofs of
           theorems and programs. *Communications of the ACM*, 22(5):271–280, 1979.

[MNPS91]   D. Miller, G. Nadathur, F. Pfenning, and A. Scedrov. Uniform proofs as a
           foundation for logic programming. *Annals of Pure and Applied Logic*, 51:125–
           157, 1991.

[Pau84]   L. C. Paulson. Verifying the unification algorithm in LCF. Technical Re-
           port 50, University of Cambridge Computer Laboratory, 1984.

[Pau87]   L. C. Paulson. *Logic and Computation: Interactive Proof with Cambridge
           LCF*. Cambridge University Press, Cambridge, England, 1987.

[Pau90]   L. Paulson. Isabelle: The next 700 theorem provers. In P. Odifreddi, editor,
           Logic and Computer Science, pages 361–385. Academic Press, 1990.

[Pea67]   Giuseppe Peano. The principles of arithmetic, presented by a new method.
           In van Heijenoort [vH67], pages 83–97. First published 1889.

[Pét67]   R. Péter. *Recursive Functions*. Academic Press, New York, NY, 1967.

[Pfe89]   F. Pfenning. Elf: A language for logic definition and verified metaprogram-
           ming. In *4th IEEE Symposium on Logic in Computer Science*, pages 313–322,
           1989.

[PH78]   J. Paris and L. Harrington. A mathematical incompleteness in Peano arith-
           metic. In Barwise [Bar78c], pages 1133–1142.

[Pit70]   J. Pitrat. Heuristic interest of using metatheorems. In *Symposium on Automatic Demonstration*, pages 194–206. Springer-Verlag, Berlin, 1970.

[Pos21]   E. L. Post. Introduction to a general theory of elementary propositions. *American Journal of Mathematics*, 43:163–185, 1921. Reprinted in [vH67, pages 264–283].

[Qua90]   A. W. Quaife. *Automated Development of Fundamental Mathematical Theories*. PhD thesis, University of California, Berkeley, 1990.

[Rob65]   J. A. Robinson. A machine-oriented logic based on the resolution principle. *JACM*, 12(1):23–41, 1965.

[Ros36]   J. B. Rosser. Extensions of some theorems of Gödel and Church. *Journal of Symbolic Logic*, 1:87–91, 1936. Reprinted in [Dav65, pages 231–235].

[Ros83]   J. B. Rosser. Highlights of the history of lambda-calculus. In *Proceedings of the 1983 ACM Symposium on Lisp and Functional Programming*, New York, NY, 1983. ACM.

[Rus67]   Bertrand Russell. Letter to Frege. In van Heijenoort [vH67], pages 124–125. Written 1902.

[Rus83]   D. M. Russinoff. A mechanical proof of Wilson's theorem. Master's thesis, Department of Computer Sciences, University of Texas at Austin, 1983.

[Rus88]   D. M. Russinoff. A mechanical proof of quadratic reciprocity. *Journal of Automated Reasoning*, 8(1):3–21, 1988.

[RvH91]   J. M. Rushby and F. von Henke. Formal verification of the interactive convergence clock synchronization algorithm using EHDM. Technical Report SRI-CSL-89-3R, Computer Science Laboratory, SRI International, Menlo Park, CA, February 1989 (Revised August 1991). Also available as NASA Contractor Report 4239.

[RvHO91]   J. M. Rushby, F. von Henke, and S. Owre. An introduction to formal specification and verification using EHDM. Technical Report SRI-CSL-91-2, Computer Science Laboratory, SRI International, Menlo Park, CA, February 1991.

[Sha85]   N. Shankar. Towards mechanical metamathematics. *Journal of Automated Reasoning*, 1(4):407–434, 1985.

[Sha88a]   N. Shankar. A mechanical proof of the Church-Rosser theorem. *Journal of the ACM*, 35(3):475–522, 1988.

[Sha88b]   S. G. Shanker, editor. *Gödel's Theorem in Focus*. Routledge, 1988.

[Sho67]   J. R. Shoenfield. *Mathematical Logic*. Addison-Wesley, Reading, MA, 1967.

[Smo78]   C. Smorynski. The incompleteness theorems. In Barwise [Bar78c], pages 821–865.

[Smu83]   R. M. Smullyan. *5000 B. C. and Other Philosophical Fantasies.* St. Martin's Press, New York, NY, 1983.

[Smu92]   R. M. Smullyan. *Gödel's Incompleteness Theorems*, volume 19 of *Oxford Logic Guides.* Oxford University Press, New York, NY, 1992.

[Ste72]   S. Stenlund. *Combinators, λ-terms and Proof Theory.* D. Reidel, Dordrecht, Holland, 1972.

[Sti88]   M. E. Stickel. A Prolog technology theorem prover: Implementation by an extended Prolog compiler. *Journal of Automated Reasoning*, 4(4):353–380, December 1988.

[Sza69]   M. E. Szabo, editor. *The Collected Papers of Gerhard Gentzen.* North-Holland, 1969.

[Tar83]   A. Tarski. The concept of truth in formalized languages. In J. Corcoran, editor, *Logic, Semantics, Metamathematics (2nd edition)*, pages 152–278. Hackett, Indianapolis, 1983. Translated by J. H. Woodger.

[Tur65]   A. M. Turing. On computable numbers, with an application to the Entscheidungsproblem. In Davis [Dav65], pages 116–154. First published 1937.

[TvD88]   A. Troelstra and D. van Dalen. *Constructivity in Mathematics.* North-Holland, Amsterdam, 1988.

[Tym79]   T. Tymoczko. The four-color problem and its philosophical significance. *The Journal of Philosophy*, LXXVI(2):57–83, February 1979.

[vBJ79]   L. S. van Benthem Jutting. Checking Landau's 'Grundlagen' in the Automath system. Technical report, Mathematical Centre, Amsterdam, 1979. Mathematical Centre Tracts.

[vH67]   J. van Heijenoort, editor. *From Frege to Gödel: A Sourcebook of Mathematical Logic, 1879–1931.* Harvard University Press, Cambridge, MA, 1967.

[vN61]   J. von Neumann. *John von Neumann, Collected Works, Volume V.* Pergamon Press, Oxford, 1961.

[Whi58]   A. N. Whitehead. *An Introduction to Mathematics.* Oxford University Press, New York, NY, 1958.

[WR25]   A. N. Whitehead and B. Russell. *Principia Mathematica.* Cambridge University Press, Cambridge, 1925.

[WT74]   R. W. Weyhrauch and A. J. Thomas. FOL: A proof checker for first order logic. Technical Report AIM-235, Stanford University, Computer Science Department, Artificial Intelligence Laboratory, 1974.

Index

Printed in the United States
By Bookmasters